AF903709

MY FAMILY ALBUM

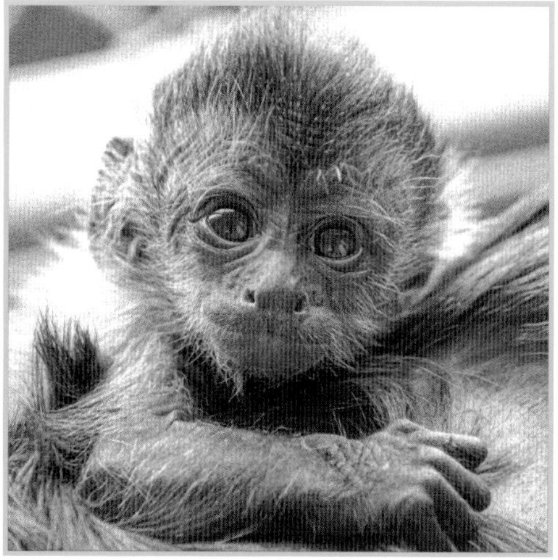

Also by Frans de Waal

Minds I've Met (2026)
Different (2022)
Mama's Last Hug (2019)
Are We Smart Enough to Know How Smart Animals Are? (2016)
The Bonobo and the Atheist (2013)
The Age of Empathy (2009)
Primates and Philosophers (2006)
Our Inner Ape (2005)
My Family Album (2003)
The Ape and the Sushi Master (2001)
Bonobo: The Forgotten Ape (1997)
Good Natured (1996)
Peacemaking Among Primates (1989)
Chimpanzee Politics (1982)

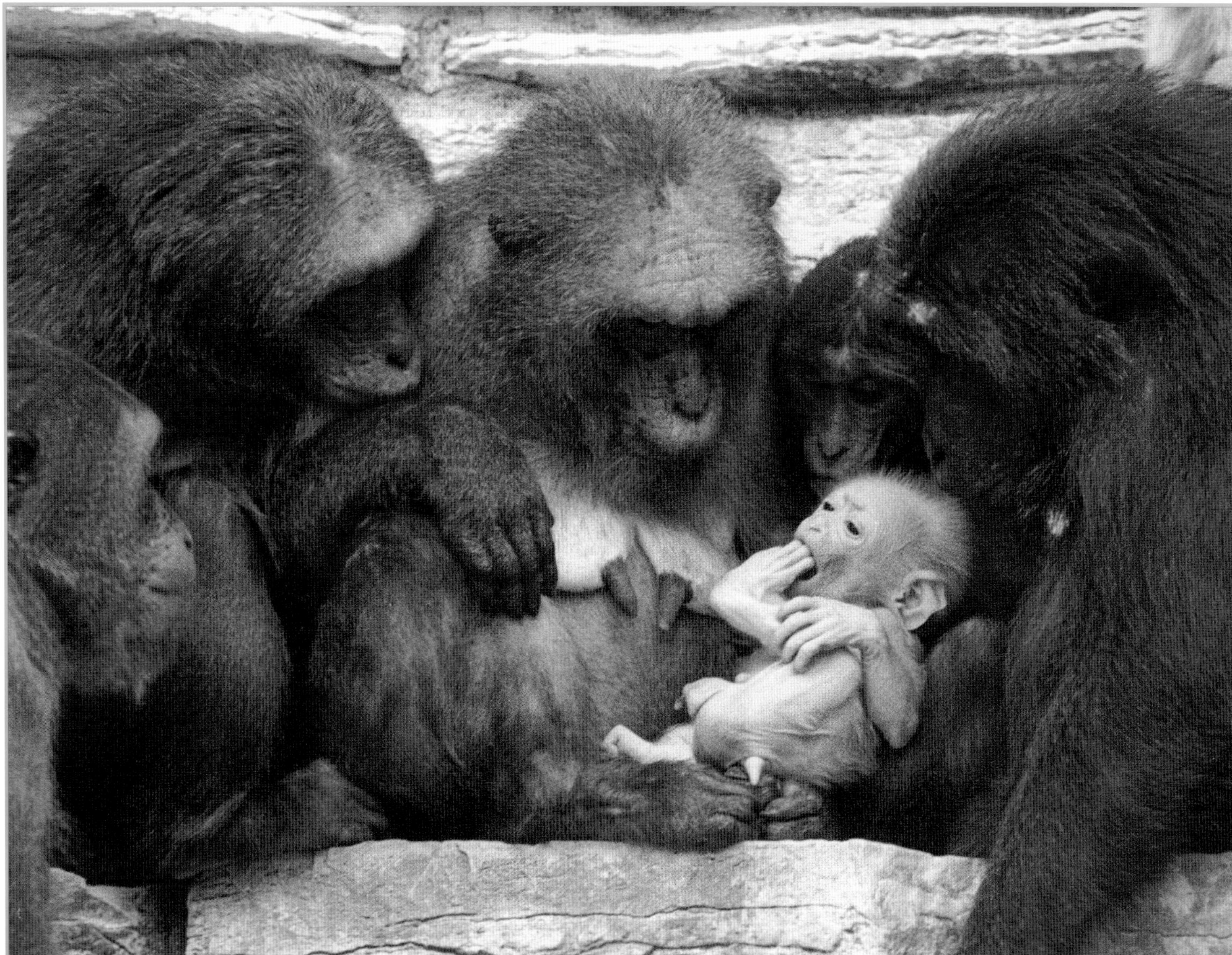

My Family Album

Thirty Years of Primate Photography

Text and Photographs by Frans de Waal

With a Foreword by Carl Safina

W. W. NORTON & COMPANY
Independent Publishers Since 1923

Copyright © 2026 by Frans de Waal
Foreword copyright © 2026 by Carl Safina

All rights reserved
Printed in China
First Edition

No portion of this work is authorized for reproduction, storage in a retrieval system, or transmission by any means in any form, and all text, images, and data herein contained are expressly excluded from use in any machine learning, artificial intelligence training, or data mining activity, absent prior written permission. The publisher waives no rights under domestic or international law and expressly reserves this work from the text and data mining exception pursuant to Article 4(3) of the European Union Digital Single Market Directive 2019\790.

For information about permission to reproduce selections from this book, write to Permissions, W. W. Norton & Company, Inc., 500 Fifth Avenue, New York, NY 10110

For information about special discounts for bulk purchases, please contact W. W. Norton Special Sales at specialsales@wwnorton.com or 800-233-4830

Manufacturing by Toppan Leefung
Production manager: Julia Druskin

ISBN 978-1-324-11784-1

W. W. Norton & Company, Inc., 500 Fifth Avenue, New York, NY 10110
www.wwnorton.com

W. W. Norton & Company Ltd., 15 Carlisle Street, London W1D 3BS

Authorized EU representative: EAS, Mustamäe tee 50, 10621 Tallinn, Estonia

1 2 3 4 5 6 7 8 9 0

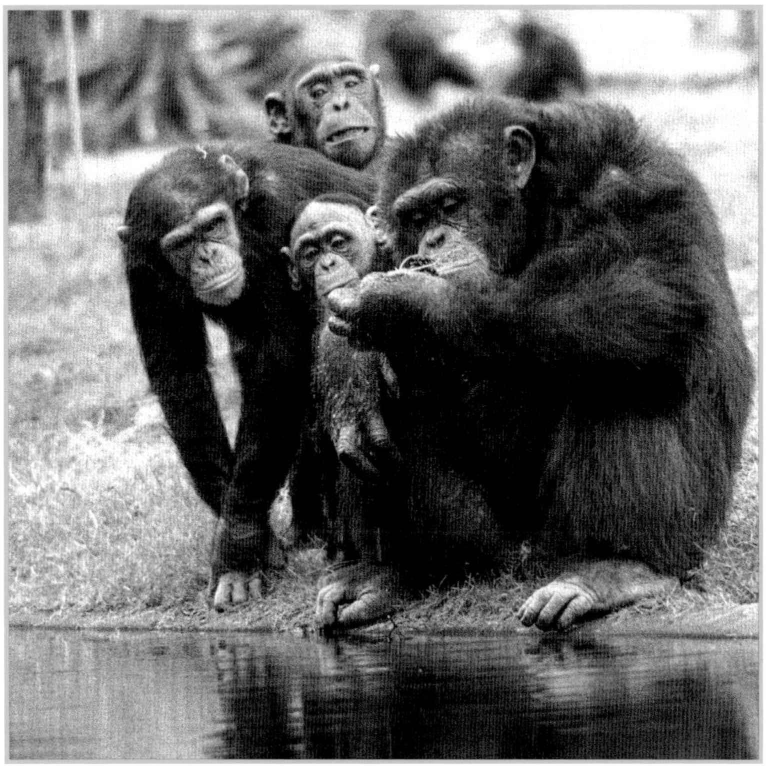

Contents

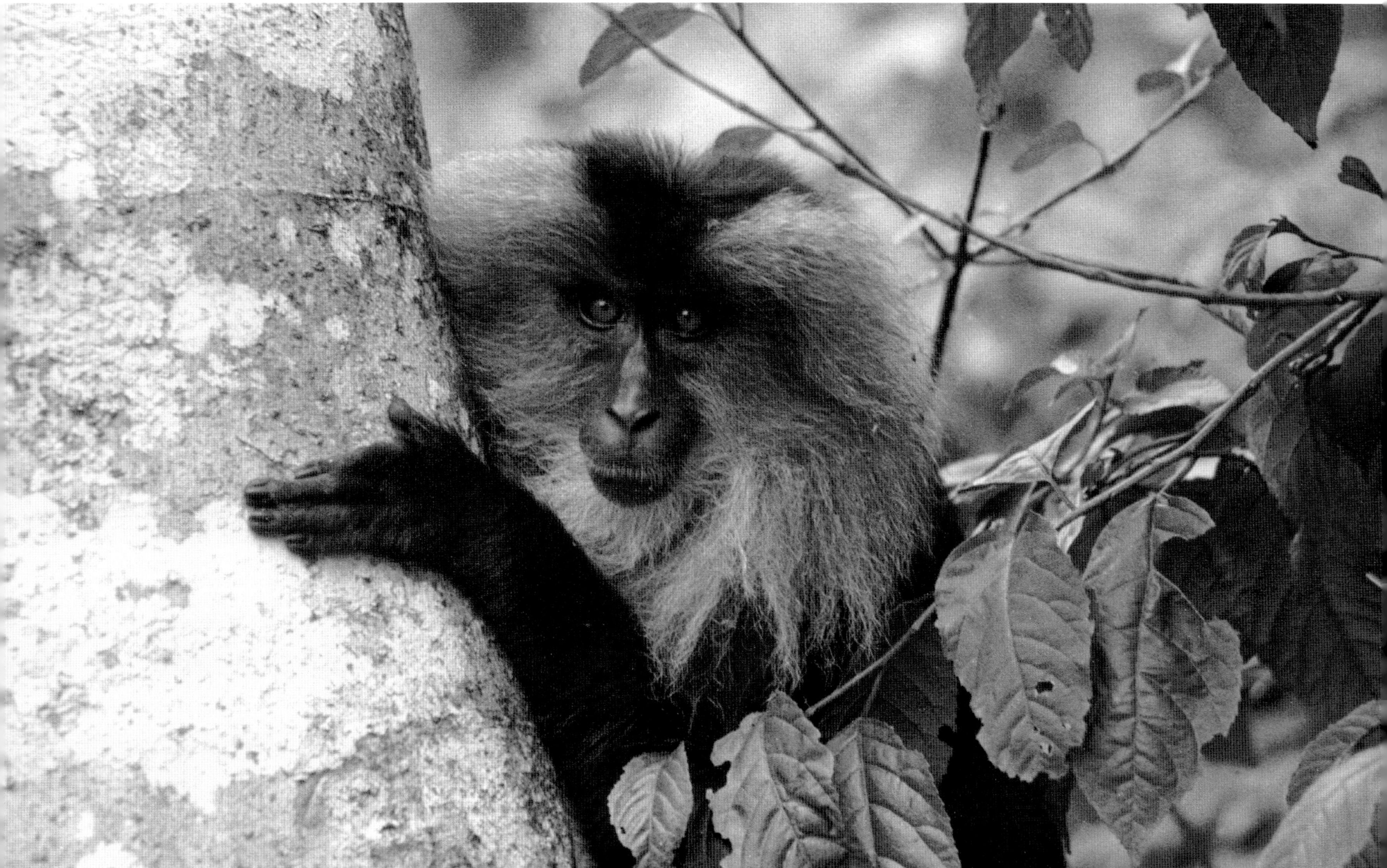

Foreword *by Carl Safina*

Frans de Waal was a bit of a stealth radical. At first implicitly and later explicitly, he built on his observations—initially shared by almost no one in his field—that empathy was not uniquely human. That empathy is in fact widespread in the other-than-human world. The implications were groundbreaking and earthshaking. But de Waal was a subtle man, and eventually his understated delivery landed this message all the more powerfully.

De Waal delivered this radical message by acting like a scientist in the very effective camouflage of actually being a scientist, and a very good one at that. Over decades, he continued publishing actively in peer-reviewed science journals. He prominently held an endowed professorship at a highly respected U.S. university. He garnered a lengthy list of scientific awards. And he was an elected member of the Royal Netherlands Academy of Arts and Sciences and the U. S. National Academy of Sciences—one of the world's highest scientific honors.

Eventually de Waal's observations and views widened to show—quite unequivocally—that a sense of fairness and the resulting morality are not strictly human. They are more widespread than we have been taught, and indeed more widespread than we ever thought. Perhaps the most "radical" and relevant message that de Waal conveyed from primates to us was that morality comes to humans from within. It is part of our human nature, derived from our evolved primate lineage; morality is not imposed by a god, nor does it require religion. This was radical not because it was far-fetched but because it lay bare our accustomed self-delusions and self-aggrandizement.

Frans de Waal had that thing that scientists most covet: ideas. And the courage to be original. His superpower—what he really excelled at—was the ability to watch animals and to see, with insight, things everyone before him had missed. Either they didn't have his talent for watching or the blinders of earlier dogma prevented them from seeing. But perhaps it's accurate to say that de Waal's ability to remove the blinders was indeed his talent for seeing.

The observed facts added up to a conviction that became de Waal's message and mission: to show the world that the boundaries we've imposed between people and "animals" are false. That the imagined boundaries are in fact self-imposed illusions and intellectual delusions. The differences are actually gradients, more like "a gently sloping beach." The reality is life's continuity. The evidence includes proximity. Other animals are closer to us than we have been taught. They are our relatives. You can see it most easily in similarities we share. Ambitions, agendas, alliances, hostilities . . . And perhaps most consequentially, de Waal observed that the inevitable and too-familiar frictions of living in groups require skilled peacemaking—reconciliation and the ability to make up and move on.

De Waal's many books include *Chimpanzee Politics* and *Peacemaking Among Primates.* You get the idea. Prior less meticulous observers and authors hyped impressions of drama and violence; those things are so obvious when you watch primates that they can distract you into false impressions. De Waal showed how much more time primates actually spend in subtle, nuanced, and skilled interactions such as bonding, cooperation, sharing, empathy, and consolation. Not just primate ambition and conflict, but conflict *resolution*. This was revolutionary.

All of these things, the bad and the good, are recognizable to us because they are familiar *in* us. We're different, but we're not so different. Central to de Waal's message, then: We are family. And thus, this book, *My Family Album.*

The title sums it up. De Waal's photos say all. But as well known as he was for chronicling and interpreting primate social dynamics in prose, Frans de Waal was not well known for photography. Indeed, he'd teamed up with the preeminent wildlife photographer Frans Lanting for the 1997 book *Bonobo: The Forgotten Ape.* That book essentially called previously obscure apes out of their green rainforest shadows and debuted them to the wider public.

Bonobos are prominent here in *My Family Album,* too. This singular relative of ours, living in female-centered, egalitarian, peaceful communities, has a thing or two to say to us. While chimpanzees live in male-dominated groups where rank is sought and fought by violent male aggression, bonobos meanwhile—equally close relatives of ours—use female dominance to suppress male aggression and keep the peace. It's something we should bear in mind.

And who can miss that the many fluid combinations of bonobo sex—portrayed in this book through de Waal's stunningly intimate, humanly recognizable photos—invite all of us to relax about our bodies and about same-sex love, and to see instead the higher need and importance of forming emotional bonds. And, of course, to "make love" (as in bonobos), "not war" (as in chimps) is a message for all of us about an evolutionary road less traveled. Among bonobos, sex is always a positive, peacemaking force—in any combination and any position. Their approach remains an option for that enigmatic primate in the mirror who, more than any, can consider our choices and decide our path.

My Family Album isn't solely about apes but about primates more widely and the familial bonds they share with themselves and with us.

Frans de Waal was born in 1948 in the Netherlands. He studied under noted primate biologist Jan van Hooff and got his PhD in 1977 for studies of aggression in macaque monkeys. By the time of his PhD he'd already begun a long and—it would turn out—highly productive study of social behavior in a large captive population of chimpanzees in Arnhem Zoo in the Netherlands.

Studying captive beings has several major limitations. Mainly, their lives are not natural. They cannot forage naturally, cannot change or leave groups on their own accord, cannot interact freely with neighboring communities as in the wild, and do not, in sum, have full agency over their lives and the freedom to express their full range of natural behaviors and psychological dynamics. On the other hand, familiarity and ease of observation confer certain advantages to a human researcher at a zoo. So there were potential benefits but also drawbacks.

For the kinds of questions de Waal was interested in, and—this is important—the kind of researcher he was becoming—the zoo environment's benefits outweighed its drawbacks. In a sense, de Waal's questions carved out those natural behaviors of chimps that are retained and expressed in a suitably naturalistic artificial environment. In Arnhem Zoo, particularly, with the world's largest population of captive chimps, de Waal struck a rich vein of observable social dynamics. He mined these behaviors; analyzed them with the formal systematic methods of science; thought about them objectively, freely, and creatively; compared his impressions with the field of primatology's prior dogmas; and emerged with new observations and new interpretations that became new understandings. Observing primates in varied settings on five continents became his life's work. The gems of his insight became his books and public talks—and the skillfully captured photos here, in this book. We are all richer for it.

Jane Goodall's years of pioneering studies with wild chimps living in a compromised situation (she provided food to bring them into observational range, which greatly affected some of their social dynamics) and de Waal's studies of captive chimps also artificially provided with food, together facilitated valid insights into the (previously dismissed and denied) abilities of chimpanzees to think, scheme, plan, form alliances, and plot their moves for advancing dominance. In short: to have individuality, personality, and the arc of a lifetime.

But while Goodall used an intellectual jackhammer for the first assault on the brick wall of human-centered bigotry toward our living relatives, de Waal had the later luxury of seeing subtleties.

After Goodall, the wall began crumbling. But after de Waal, the evidence and the formal scientific case were undeniable. He ratcheted our wider understanding forward, permanently. After de Waal, there was no denying, no going back.

As portrayed in *My Family Album*, de Waal's interests extended beyond apes. De Waal's experiments proved that monkeys, too, are perceptive enough of others to demand fair treatment. With a graduate student he expanded his attention to elephants. He came to see, and to say, that empathy was a capacity widely shared by mammals. It formed a basis in morality for them, as for us.

His early work weathered the criticism of the time. More recent work of other researchers studying primates, elephants, rodents, dogs, and others have confirmed that empathy and fairness are vastly more widespread than we'd been taught. If our favorite story about ourselves was how special and elevated we are, de Waal helped spoil it. The real story is a much better one: that we are not alone, that we are here among family, that we as human beings are related not just physically but also psychologically to other sentient, feeling companions of this planet.

Frans de Waal died in 2024. We might have thought there would be no more books from him. But he left some work in progress. And this welcome reissue of his rare and long-out-of-print *My Family Album* is overdue.

In the hands of a master observer such as de Waal, a camera becomes a tool of art and of personal intimacy. For the viewer such as you and me, it's a tool for seeing into the soul of the photographer as well as the minds of those individual primates who are photographed here. All live on in these photos. They show us that we were never alone here on this miraculous planet. And that there are other lives at stake.

Introduction

Being around monkeys and apes every day, I am fascinated by the richness of their interactions. I follow their social lives the way people follow soap operas. Who loves whom, who hates whom? What happened yesterday, what will happen today? These animals are family, gossiped about on an individual basis, personalities and all. For the nonprimatologist they are family, too—original family, in fact—if not literally then at least in the larger scheme of people's relation with nature. This closeness is best confirmed through pictures: hence this album for the enjoyment and edification of those who usually see primates only a few minutes at a time, for instance, at the zoo. People rarely get to see much of primates' social lives, which is exactly what I try to show in the ensuing pages.

Primate Photography

Even though humans are primates, taking pictures of other primates is totally different from taking pictures of people. A photographer can ask people to stand still, look at the camera, or act in a particular way, none of which works with untrained monkeys and apes. This, combined with the fact that the most interesting and exciting moments in a primate's life are full of bustle, makes me feel more like an action sports photographer than a portraitist. Apes and monkeys are fast-moving objects with rapidly changing facial expressions, postures, and positions. They run or climb around at superhuman speed. The photographer's challenge is to immobilize this mobility. The end result is a tranquil-looking image showing an outstretched hand, a grinning face, or a loving touch, but none of these now frozen moments has actually lasted for more than a fraction of a second.

Apart from collecting such situations, the task of the photographer is to present the primates in a dignified fashion. They deserve to be depicted in a manner that fills the human viewer with both curiosity and respect. The Chinese proverb that a picture is worth a thousand words definitely applies here. One can explain over and over how much we share with, say, a bonobo, but it is only when we look the ape in the eye, either live or in a two-dimensional representation, that the similarity hits home. Pictures of our close relatives reach recesses inside us that we often don't wish to be reached, given the dogma that we humans are the planet's only intelligent life form. Images are harder to ignore than words. I believe that our evolutionary heritage becomes undeniable in the face of animals with instantly familiar postures, hands, glances, and gestures.

My primary goal has been to capture *social* moments, such as tense confrontations, happy reunions, playful teasing, sexual intercourse, mother-child care, and the like. I thus seek to illustrate what primate group life is made of. I try to catch in a single frame, for example, the faces of two individuals meeting. Or I set out to show how the loser of a fight bares his teeth in a nervous grin. Picking such moments out of the stream of action that unfolds before the observer's eyes requires anticipation, which in turn requires familiarity. Since my job of the last thirty years has been the study of primate behavior, I am very much in tune with their sociality, ready to press the button at the right time. The result is a family album that reveals the family's ups and downs along with the eternal cycle of infancy, youth, adulthood, and old age.

Besides the obvious professional reasons for my photography, there is also my lifelong interest in the visual arts. I used to draw and paint, and at one point I considered becoming an artist rather than a scientist. I discovered photography in my early twenties, learning much from my then-girlfriend, now wife, Catherine, who is a talented (human) portrait photographer. She suggested that I start out with black and white in order to develop a critical sense of focus, shadow, and composition before moving on to color, which is considerably more forgiving. Catherine also taught me to never be satisfied with the result, to recognize that it always can be improved upon.

I never moved much beyond black-and-white photography, however, and still now shoot two black-and-white rolls of film for every roll of color. The constraints of black and white are also its strengths. In the same way that a forcefully drawn figure or landscape—as in the sketches of Michelangelo or the etchings of Rembrandt—often touches deeper than a fleshed-out painting, black-and-white photography is perfect at conveying form and composition. Color would only distract from the structural features I wish to show, such as the lines in a face or the subtle communication between individuals.

I have always worked with Minolta equipment, ranging from cameras with manual focus and exposure to the fully automatic modern cameras. My lenses range from 50 to 500 mm, but since I do not always have the time to change them or fetch a tripod when something of interest catches my eye, I often walk around with a 200–300 mm zoom. I need as much light as I can get to successfully freeze movement, and I accomplish this by pushing 400 ISO black-and-white film to 800 and sometimes 1600 ISO. This provides the flexibility to capture the fleeting kiss, jump, or bite. I am not a serial shooter; that is, I don't click away and select the best pictures after development but tend, instead, to wait for specific moments that interest me. If the event looks good through the viewfinder, I will take one to three photos. The present album is a highly selective presentation, however, given that it includes 122 photographs, or roughly one-quarter of one percent of the 50,000 frames that I estimate to have taken over the years.

My most important rule for selection has been an aesthetic one: to include only the most appealing, best-quality images with an occasional compromise for rare, surprising, or unusual scenes, such as a baboon running into a gazelle calf on the Kenyan plains or a wrestling match between two submerged monkeys in a hot spring. To add to the collection, I tried to turn some of my color photographs into monochromes but was mostly dissatisfied with the washed-out results typical of such conversions. Only a dozen of the photos in this album were originally taken on color (slide) film, and many of these are reproduced in small size.

One of my heroes, Ansel Adams, once said, "A great photograph is one that fully expresses what one feels, in the deepest sense, about what is being photographed." Even though Adams specialized in immobile objects and landscapes, and hence faced a completely different set of challenges, I look at photography in the same way. My outlook is reflected in the choice of themes for this album, which relate to my studies of power relations, cooperation, food sharing, reconciliation, facial communication, and so on. These themes have inspired me to sit and wait patiently for the right shot of a highly specific event, such as that of one chimpanzee handing food to another or providing a reassuring embrace to a distressed companion. The camera angle and framing reflect how I look at my subjects, as a fly on the wall. I have no intention or desire to participate in their social affairs, even though I empathize with them. A voyeur of intimate scenes, I'd much rather see the animals pay attention to each other than to me.

There are many photographers who seek to make primates look comical by capturing them either in strange postures or dressed in clothes and sunglasses, or that seek to make them look serene and keep them still by drugging them. Increasingly, I also see images that have been doctored on a computer. Such photography, too, conveys how the artist feels about his or her subject. Perhaps it reflects a certain discomfort with our relatives and a desire to confirm the dividing line between them and us. The goal of my photography, in contrast, is to confirm our close relationship. I attempt to convey apes and monkeys with the same dignity any photographer would accord human subjects. This is relatively easy to achieve with close-up portraits. Portraits speak for themselves, since they leave no doubt about the self-assured personalities of our fellow primates. Social scenes require more explanation but have the advantage of showing the complexity of primate lives. My text identifies species and situations so that the reader will feel like a second fly on the wall and can look in on the fun and the fights that our lively family engages in.

The Subjects

The primate order includes roughly two hundred species, divided into several families. Humans belong to the Hominoid family, which also includes the four great apes: bonobo, chimpanzee, gorilla, and orangutan. Apes are not to be confused with monkeys, which form quite separate branches on the primate tree. Humans and apes have flat chests and lack tails and are considerably larger than monkeys. Roughly one-third of the photographs in this book show chimpanzees, one-third bonobos, and the final third macaques, baboons, and capuchin monkeys. Capuchins are the only New World primates depicted. The selection of the nine species for the photographs was determined by the primates that I have worked with over the past decades.

The studies that my students and I conduct are nonintrusive. Most of the time we just watch what the primates do spontaneously, but we sometimes ask them to enter situations in which they can obtain food by working together or are given a joystick to play computer games, which is a good way of testing their intelligence. I use the word *ask* since we cannot force obedience from large animals such as chimpanzees: we are dependent on their willingness to participate when we call out their names. Nine out of ten times they are eager, especially for computer games, which they enjoy as much as human children do.

Our work is carried out in captive settings, usually on large primate groups in spacious, grassy enclosures at some of the world's more enlightened zoos and academic institutions, such as the San Diego Zoo; the Arnhem Zoo, in the Netherlands; and two of the National Primate Research Centers in the United States. As a consequence, the majority of photographs in these pages are of primates living in human care. From comparisons with primates in the field, we know about the fundamental similarities between wild and captive primates in social interaction and communication. Wild primates use the same facial expressions and have the same sort of social relationships as captive primates, even though the former obviously can travel far and wide and maintain a different time budget for activities because of their need to find food.

Regular visits to my colleagues' sites in the field have also allowed the inclusion of quite a few photos of wild primates. There is, for example, a close-up of one of the world's rarest and most striking monkeys—the lion-tailed macaque—taken in its fragmented habitat in Mudumalai National Park, in India. But there are also photos of olive baboons on the 18,000-hectare Kekopey cattle ranch, in Kenya, as well as of Japanese and Tibetan macaques in their natural habitats.

Most of the monkeys and apes photographed in captivity are extensively familiar with me, the man behind the camera. Some have known me for fifteen years or more. For example, Socko and Georgia, two chimpanzees at the Yerkes Primate Center Field Station, have seen me around since they were juveniles, and the Arnhem Zoo's oldest female, Mama, and I go back thirty years. These individuals do not turn away when I aim my lens at them, as they often do with unfamiliar photographers. In fact, we have one chimpanzee who slaps both hands over her face and dives for cover behind an object as soon as a stranger aims a camera at her. In contrast, some of the portraits in this book show individuals staring straight at me, without a drop of hesitation or suspicion, displaying a trust gained over years of dealing with each other.

Some of the photographs go back to my earliest studies, when I observed the political machinations among the chimpanzees of the Arnhem Zoo. Many of my negatives from this time are still in excellent condition. In the early 1980s, I moved to the United States, where I worked with rhesus and stump-tailed macaques at the Vilas Park Zoo, in Madison, Wisconsin. In the same period, I conducted a study on bonobos at the San Diego Zoo. The apes were kept in an artificial rock enclosure. This concrete environment, with its surrealistic "Martian" backgrounds, was great for photography, but I am glad that the bonobos have since moved to grassier enclosures.

This album is intended first of all for visual pleasure. It is not a scientific work offering a new theory or new information. There is also no story or overarching theme: the text remains entirely subservient to the photography. The reader should be able to open the book at any page, look at the image(s), and read the accompanying text. The photographs have been loosely organized according to themes (such as facial expressions or dominance interactions), but I have not emphasized these themes, which are for the reader to uncover.

I keep taking pictures of my favorite subjects, even though one would think that at some point the photographer has seen it all. I have not reached this point yet and am still occasionally surprised by a rare interaction or a behavior no one has ever described before. For example, when the chimpanzees at my latest work location, the Yerkes Primate Center Field Station, started showing hand-clasp grooming, a behavior until then observed only among wild chimpanzees, I had to get the perfect picture. Entire days would go by in which the apes would not show any hand clasping at all, but there were other days in which they kept doing it. Animal photography is a matter of having the camera ready at the right time with the right light, such as a slightly overcast sky. The odds are small and unpredictable, and many things can and will go wrong, yet the animals have a way of rewarding the persistent soul.

Plates

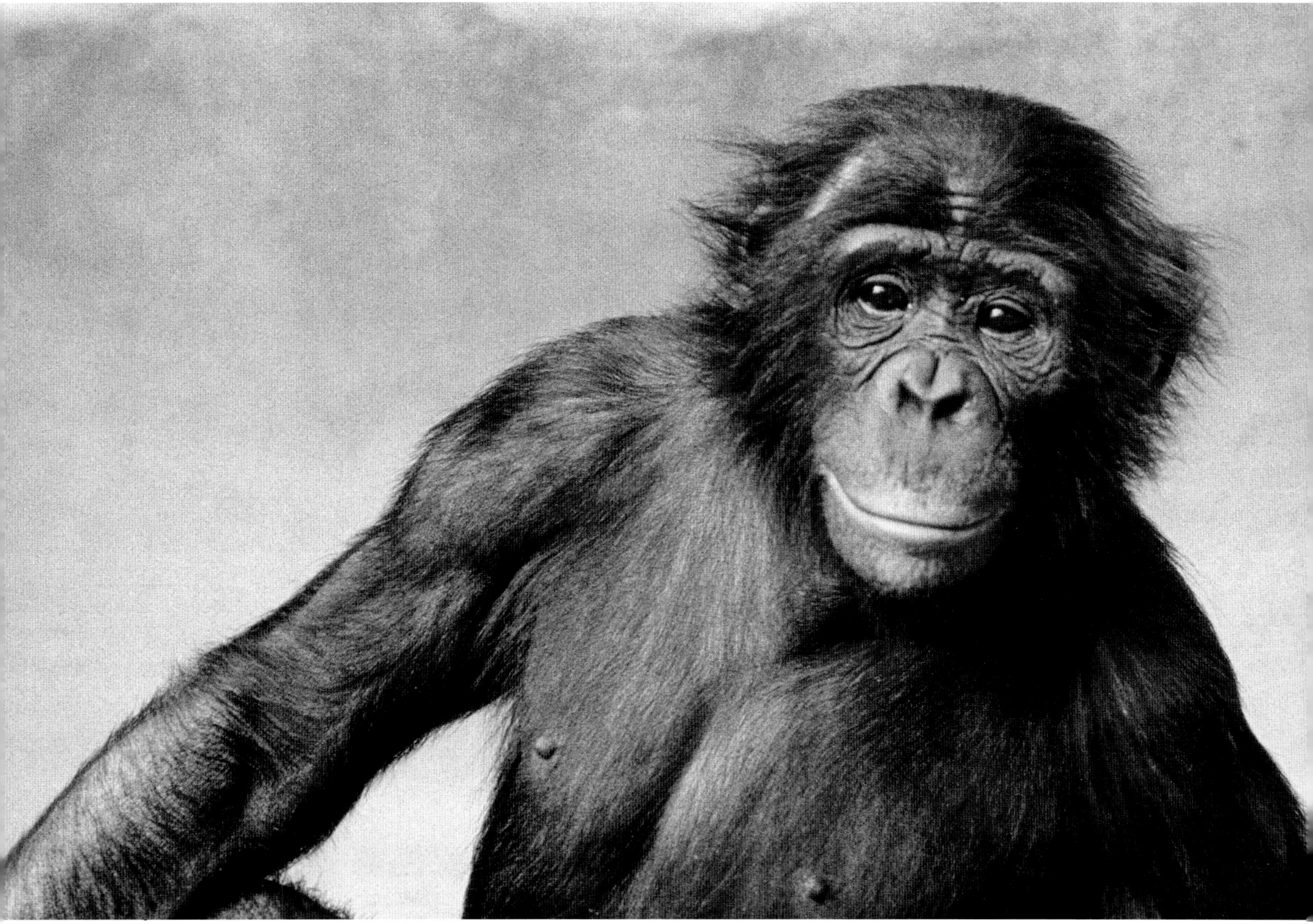

Cover boy

When we look at the face of an anthropoid ape, such as a bonobo, it becomes impossible to maintain a sharp dividing line between human and animal. The curiosity, vivacity, and self-confidence exuded by the ape leave no doubt that here we have an animal with a personality, a will, and an active mind that tries to exert control over its environment. The idea of animals as slaves of their instincts or mindless automatons evaporates. This is not just because of the physical similarity with us; it is supported by what we know about ape behavior, which is insightful, creative, and intentional.

Kalind, the younger brother of Kevin (pages 126–27), was the more excitable and playful of the two. He was so strikingly handsome that people often commented on it, and every zoo desired him as a breeding male. In this picture, Kalind is only seven years old and could still do pretty much whatever he wanted. It is only at later ages that young males have to reckon with the older, more dominant males, who increasingly place limits on their behavior.

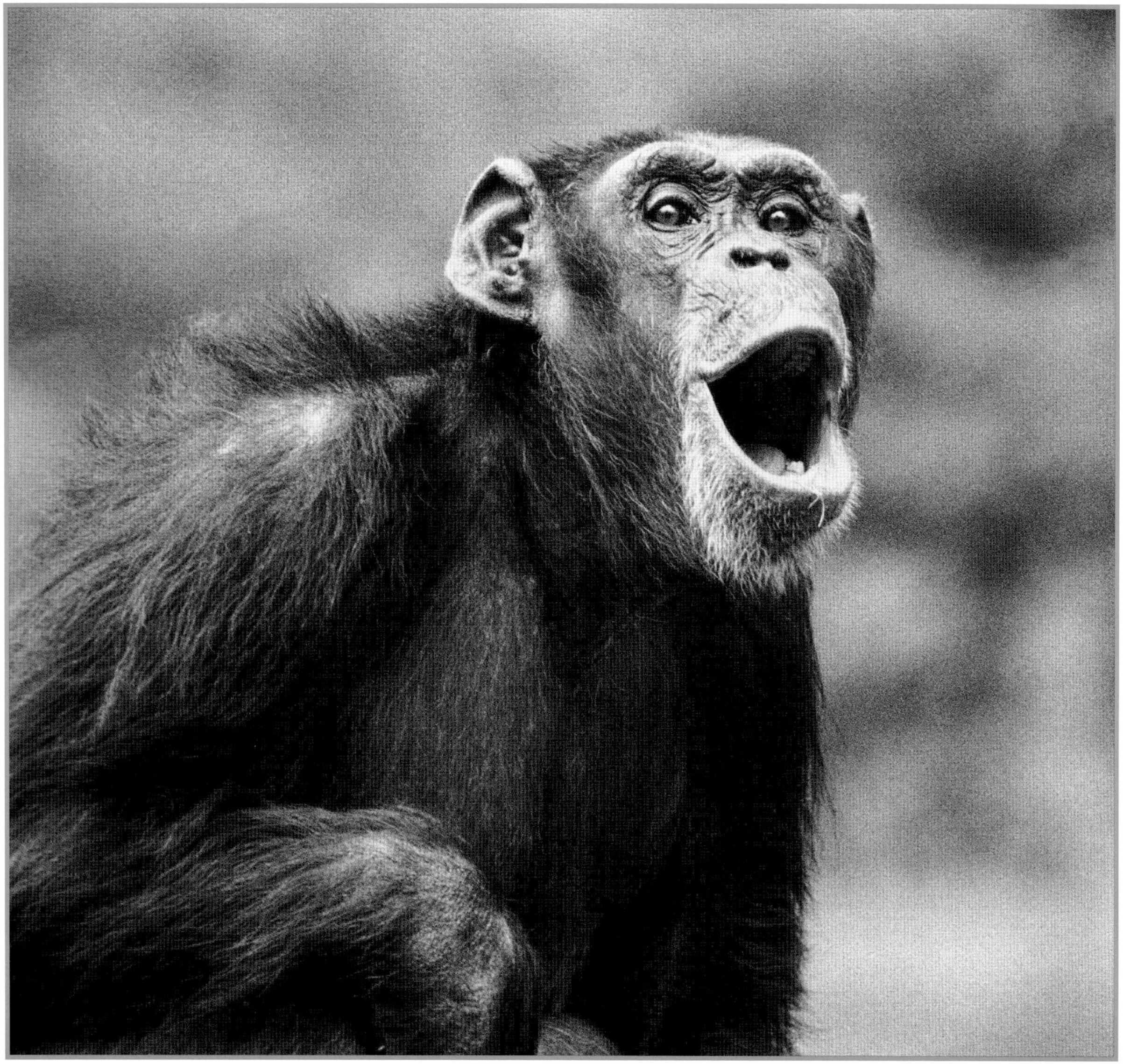

Cries of the deaf

Krom, a female chimpanzee assumed to be totally deaf, nevertheless expressed all the varied calls of her species. The keepers were the first to report her handicap, claiming that she never paid attention to any noise. Outdoors, we began to notice that she reacted to sudden sounds or alarms only after having *seen* others in her group react. For example, if she saw others run away from a charging adult male, she would run, too. She thus relied on secondhand information. A few times, I tested Krom up close in the night quarters by waiting until she had her back turned before clapping my hands and yelling: unlike the others, she never looked around. Obviously, Krom gleaned enough information from facial expressions and gestures to handle herself socially, but her disability proved lethal for her offspring. Chimpanzee infants produce noises that alert the mother to their needs. Krom lost several babies because she sat on them and did not notice or failed to nurse them when they cried for food. We ended up removing offspring from her before this point was reached: Roosje (pages 36, 142–43), for example, was placed in the care of another female.

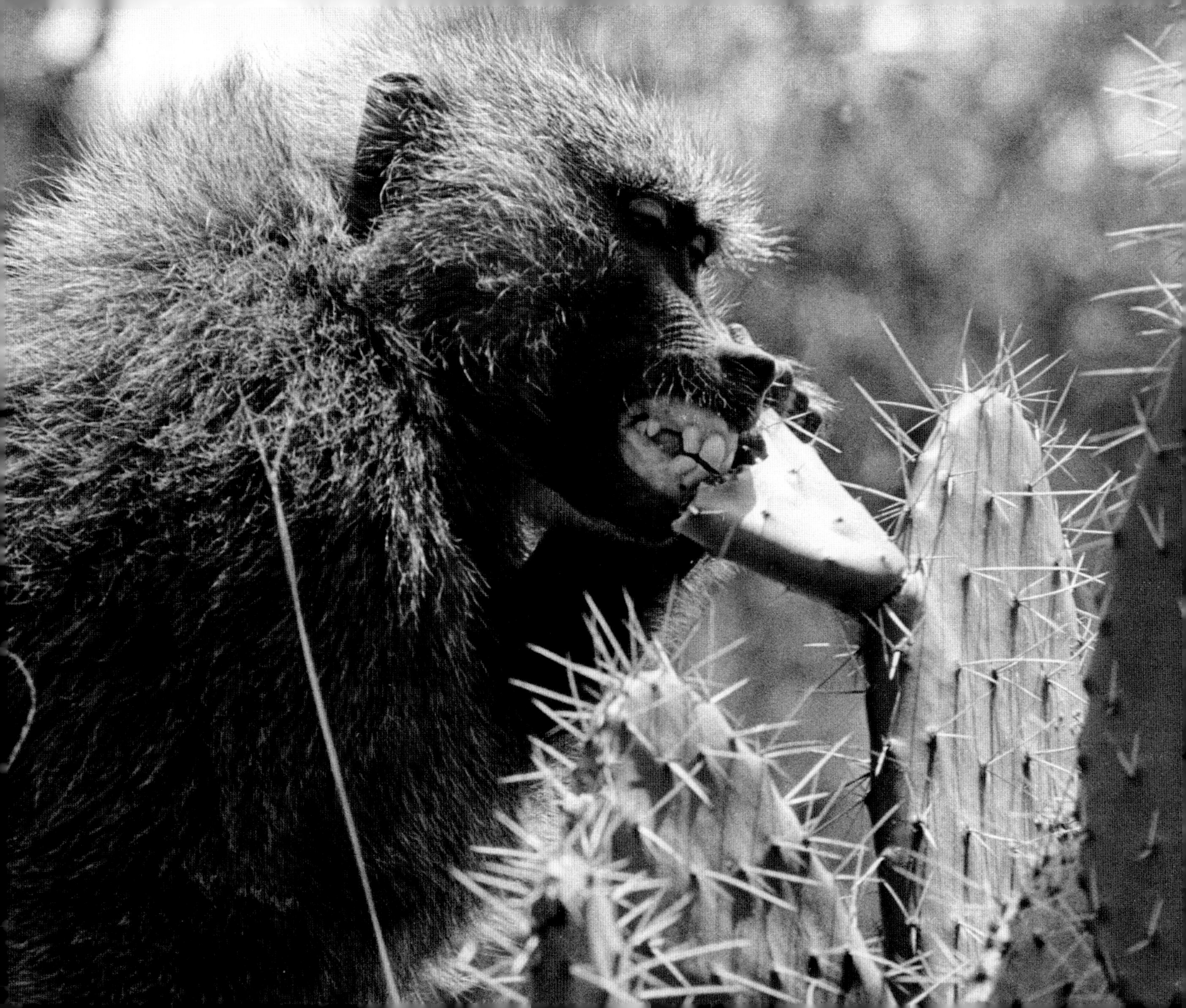

Origin of the smile

Because of an enduring drought, the olive baboons on the Kenyan plains were hungry and dehydrated. For days, they had been eating acacia beans, creating a rather smelly trail for us to follow. They occasionally lessened their thirst by eating cactus, a New World invader of Africa. I realized that this would be the perfect occasion to photograph the origin of teeth baring, a behavior debated by students of facial expressions. When this baboon tackled food that might hurt, she fully withdrew her lips. We, humans, show the same reflex, not only to prickly foods but also when peeling a lemon that sprays acid in our face. The chimpanzee in the small photo does the same to a thorny blackberry branch.

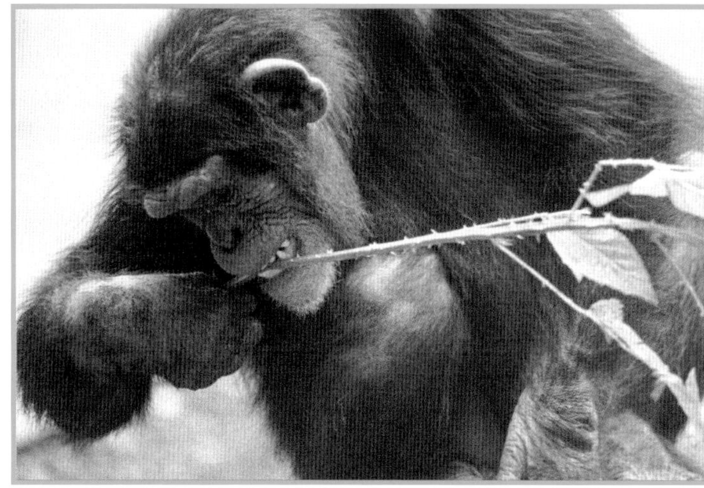

Evolution has a habit of turning reflexes into communication signals. In a process known as ritualization, a common behavior becomes more stereotypical and conspicuous so that it can be recognized from great distances. The grin and smile are considered ritualized signals derived from protective lip retraction in reaction to harmful stimuli. Many primates convey the signal to potentially harmful social partners, such as dominant individuals (pages 14–15).

Waving the white flag

Do two rows of bared teeth signal happiness? For many primates the expression means something totally different. Their grin differs from our smile. First, instead of being pulled upward, their mouth corners are pulled back and slightly down. Second, the context is different: teeth are bared when the individual feels intimidated. Thus, the Tibetan macaque male in the small photo grins in response to a threat. And the juvenile rhesus monkey in the large photo sits paralyzed while the alpha male of his troop comes near.

Like the insignia on a military uniform, this facial expression serves as a white flag to clarify social rank. Baring of teeth tells the dominant primate not to worry, that it is clear who is boss, that no one would dream of disagreeing. The same expression carries a different meaning in our species. Or does it? For humans, too, the smile has a nonthreatening, appeasing quality that sometimes betrays an element of fear. Hence people who smile a lot may be considered nervous. The original function thus remains visible, even though in our case the grin also has a friendly and happy quality.

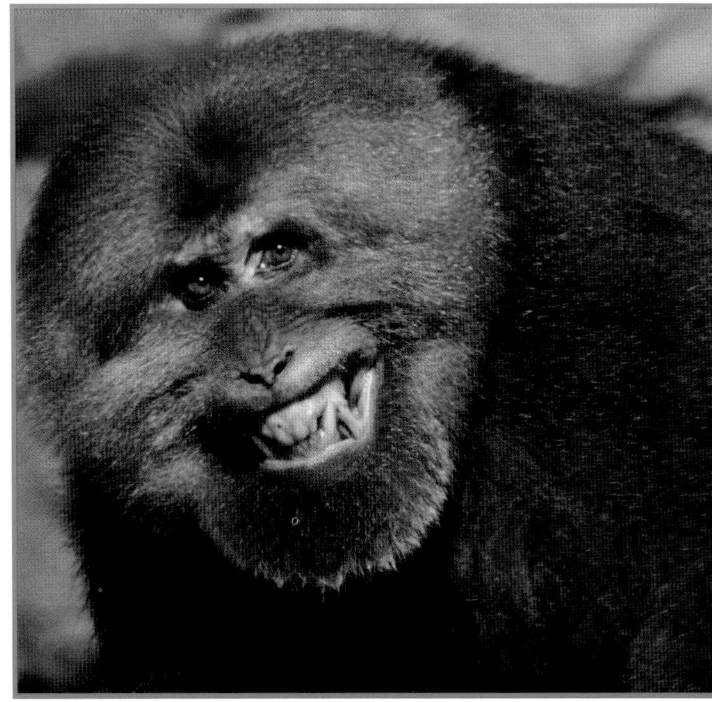

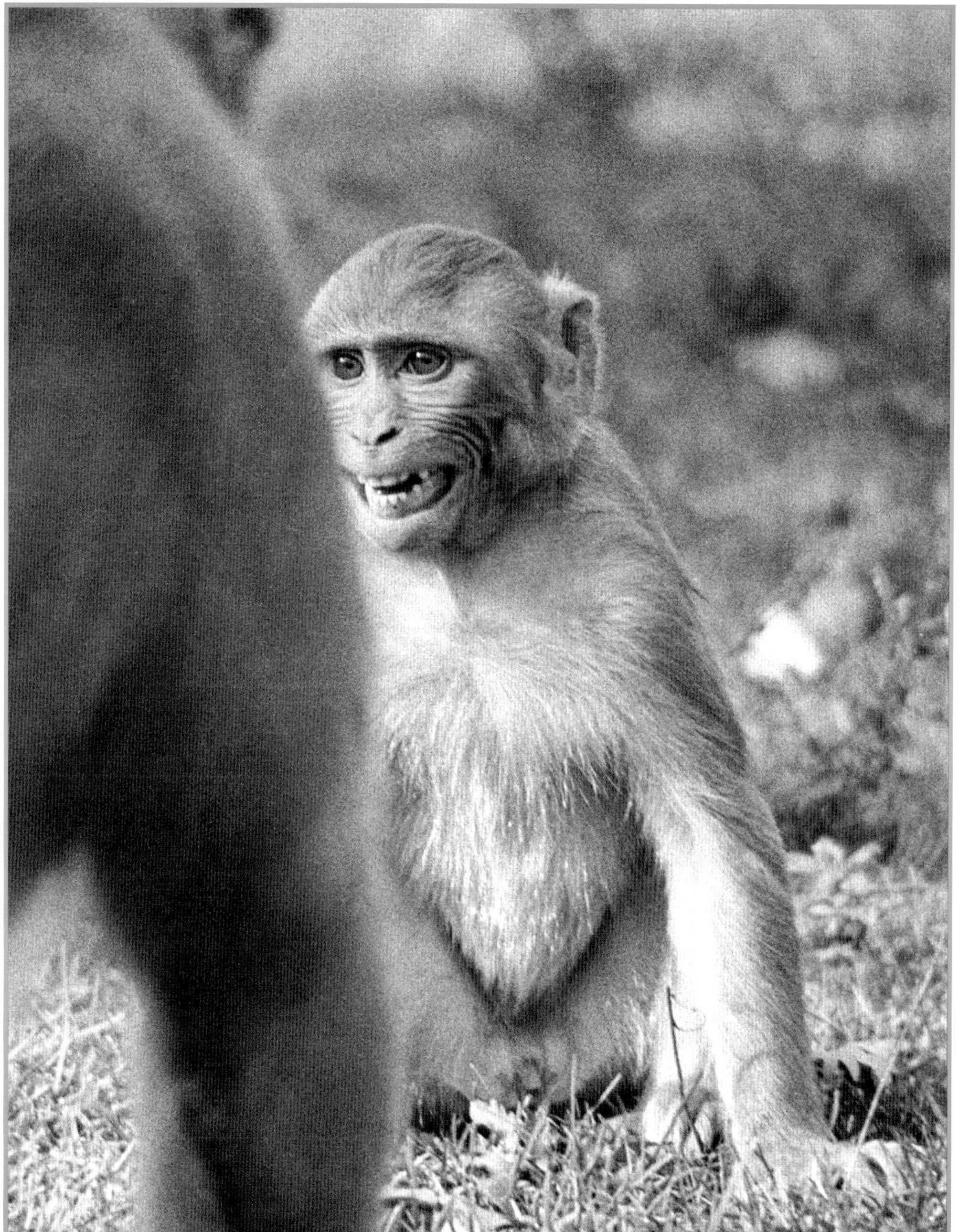

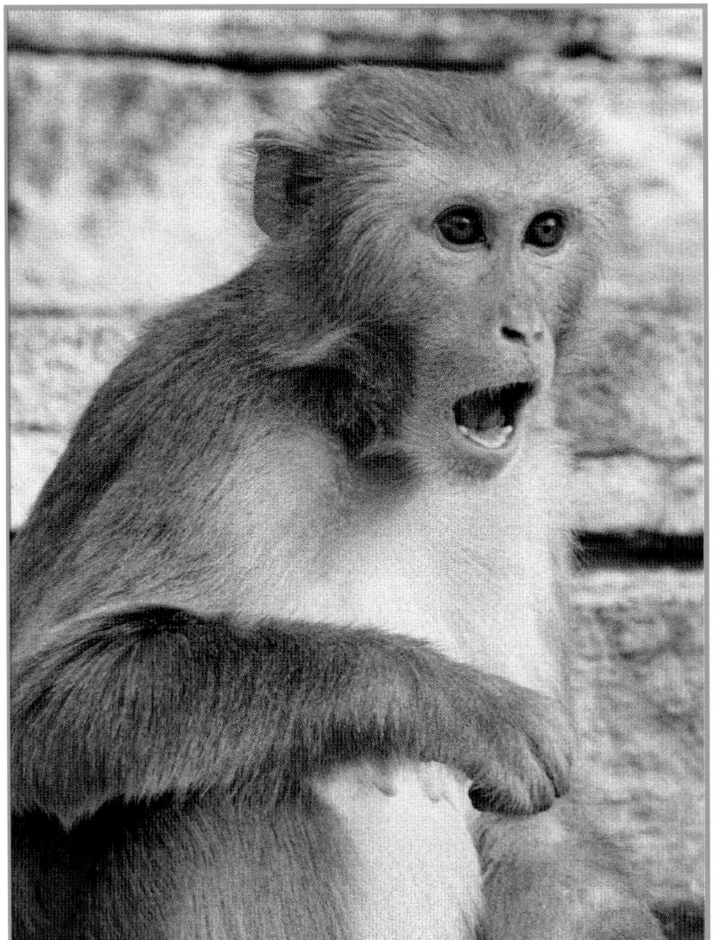

Confrontational stare

Visiting Koshima Island's Japanese macaques, my translator and I ran into an adolescent male. The monkey was by himself in the forest, away from the main troop. My companion sat down and absentmindedly put her binoculars on the ground, which the male immediately grabbed. I knew we would lose them if I didn't act right away. I normally do not confront primates, but before the monkey could even think about what to do with his new toy I stepped forward. He dropped the binoculars, and I picked them up. This little dominance struggle triggered a threatening face: the male opened his mouth and stared at my companion, who had remained seated (he obviously felt it was safer to stare at her than at me). We did not budge, and I took this picture.

The open-mouth stare is the common way in which macaques threaten each other, shown in the small photo by the alpha female of a rhesus troop. That she shows virtually no teeth in her expression means she is extremely self-confident. As soon as there is a little fear or nervousness on the part of the threatening monkey, some degree of teeth baring will occur, as is shown by this young male, who was, after all, facing an adversary many times his size.

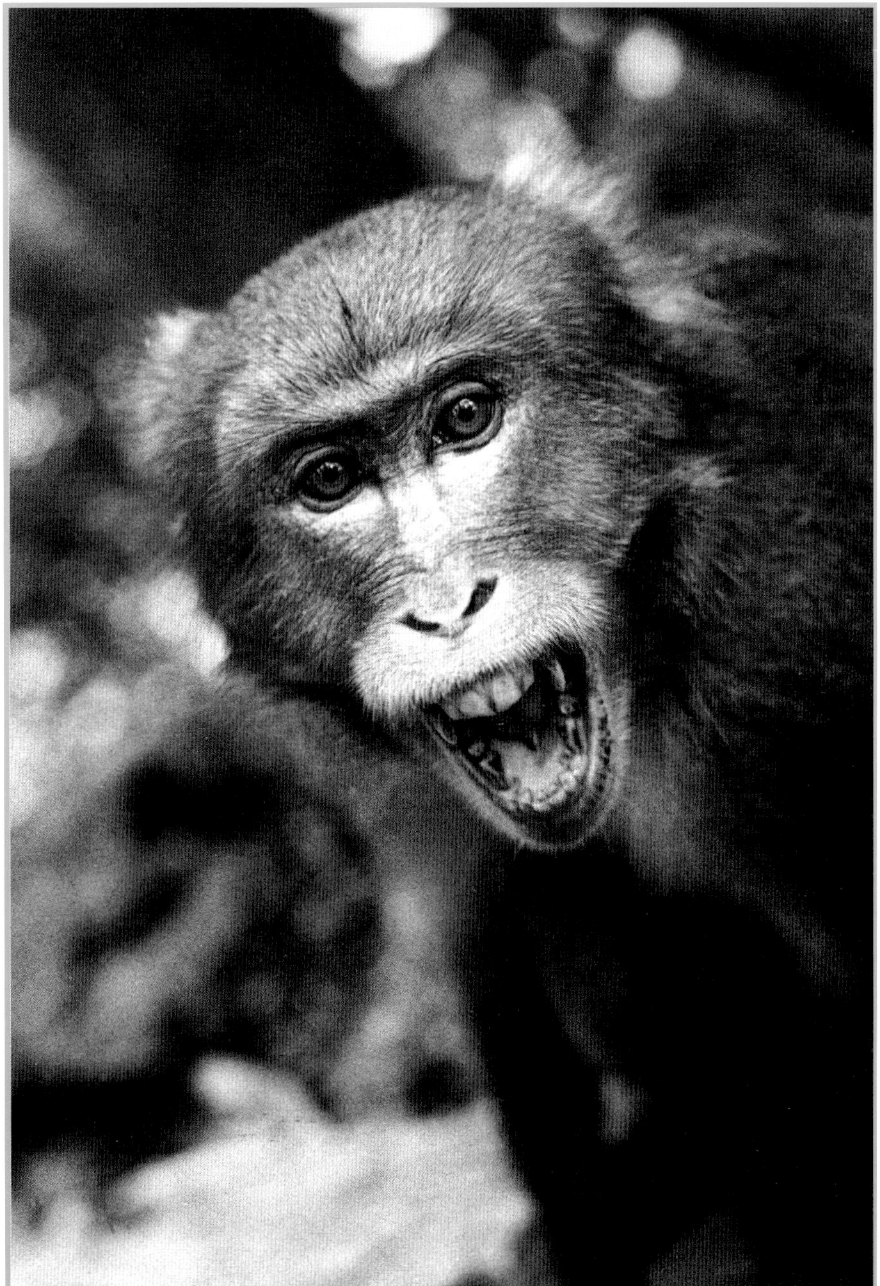

Sense of humor

Human laughter derives from the primate's "play face." Not only do the human and ape expressions look alike—with half-open mouth and relaxed muscles around the eyes—the accompanying sounds, too, have much in common. In bonobos, laughter is a hoarse, rhythmic breathy sound heard especially during intense tickling matches. In the large photo, a juvenile bonobo shows the "classic" play face with the upper teeth covered. The adolescent bonobo in the small photo illustrates a more intense form, with partially bared teeth, together with loud laughter, in reaction to an adult male who is tickling his tummy.

In the same way that people burst into laughter at an illogical or unexpected turn in a story, ape games involve incongruities, such as a youngster wrestling down an adult, or unanticipated events, such as a sudden reversal in who chases whom. Admittedly, what causes apes to chortle is rather physical, hence quite removed from our species' refined sense of humor; yet if one sees an infant chase the alpha male, who flees laughing, it is clear that their amusements share a thing or two with human jokes.

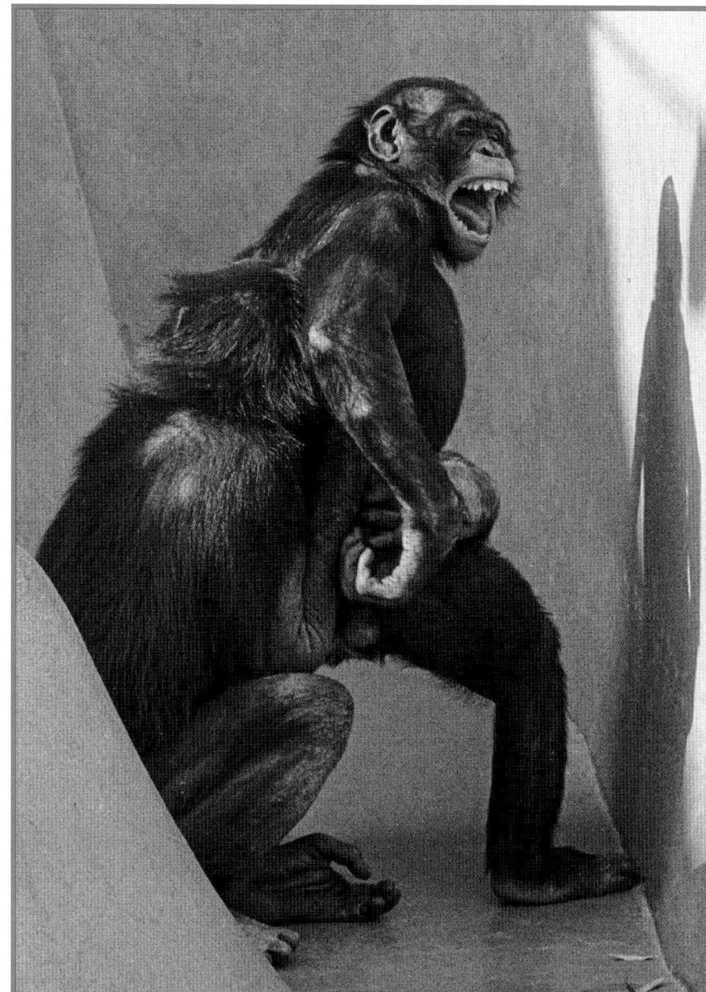

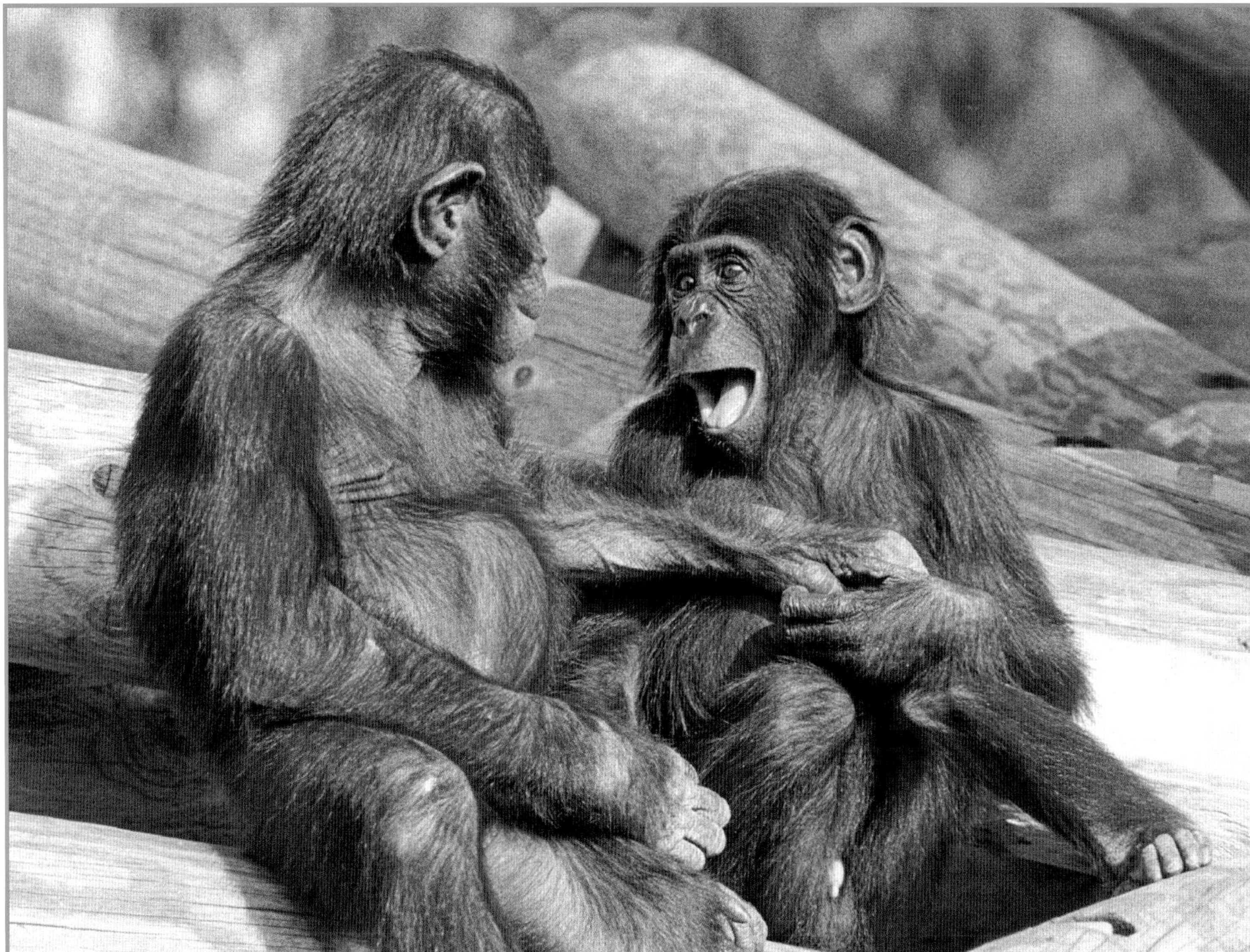

Eye contact

A male bonobo is looking straight and steadily into the eyes of his female companion while a low-level playful look spreads over his face. This expression, which most of the time serves to communicate playful intentions, can also just signal a pleasant contact. It thus serves as a smile. There is no frowning, teeth baring, or other sign of tension in the male's face, only pure friendliness.

Direct, unwavering eye contact in relaxed situations is what makes apes seem so human to us. It gives a special jolt if the eye contact is directed across the species barrier; no one can look an ape in the eyes for any length of time and maintain a sense of our species' uniqueness. In monkeys, on the other hand, rules for eye contact are different: since staring signals hostile intentions, monkeys avoid looking each other in the eyes *particularly* when they are friendly, in order to reduce the chances of misunderstandings.

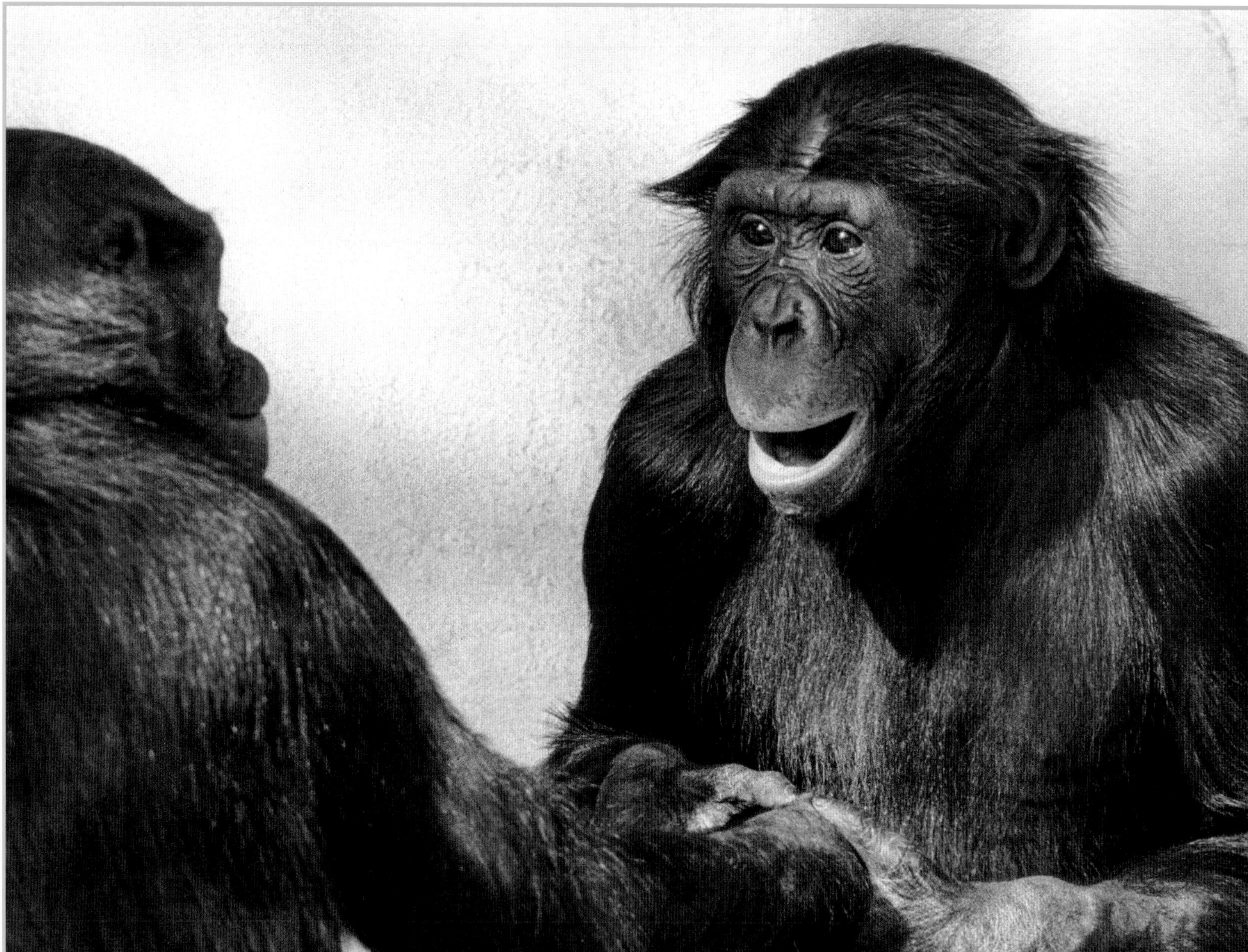

Tickle-me chimp

For close-ups of juvenile chimpanzees I went to the Yerkes Primate Center nursery, where I was let into the playpen of a rambunctious band of three-year-olds. At first they were intimidated, but they settled down as soon as I had made a few "jokes." In chimpanzee terms, a joke consists of acting as if you are about to grab someone (and missing on purpose) or playing peekaboo from a hiding place. A routine of running, stumbling, and falling is also funny to them. Ape humor is at a slapstick level. Once I had the juveniles in the right mood, I ended up getting a shot by holding my camera in one hand while using the other hand to poke a young male in the belly. Chimpanzees have the same ticklish spots as we do (the belly, sides, armpits, neck), and they utter a hoarse, guttural laugh. When the tickling gets intense, they almost choke on their guffaws. The expression in the large photograph is at top intensity, equivalent to people roaring with laughter. The small photo shows a relaxed portrait of the same male, before I had tickled him.

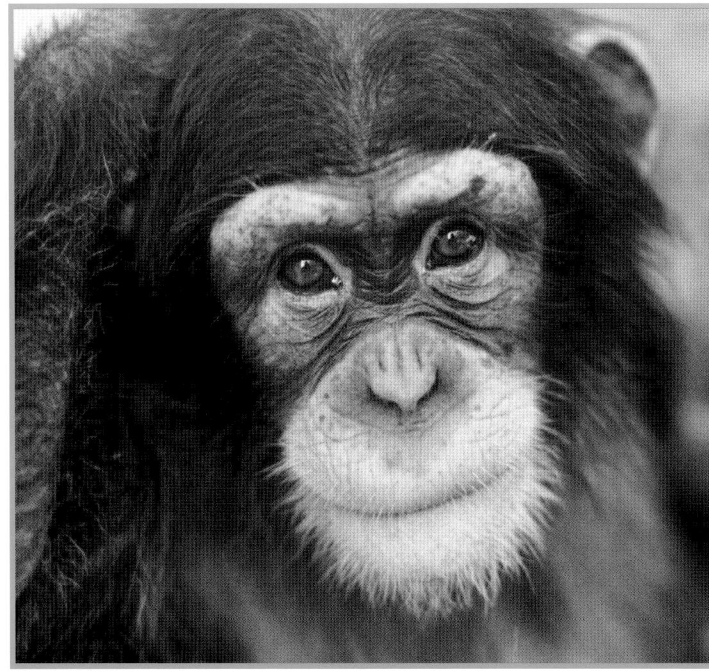

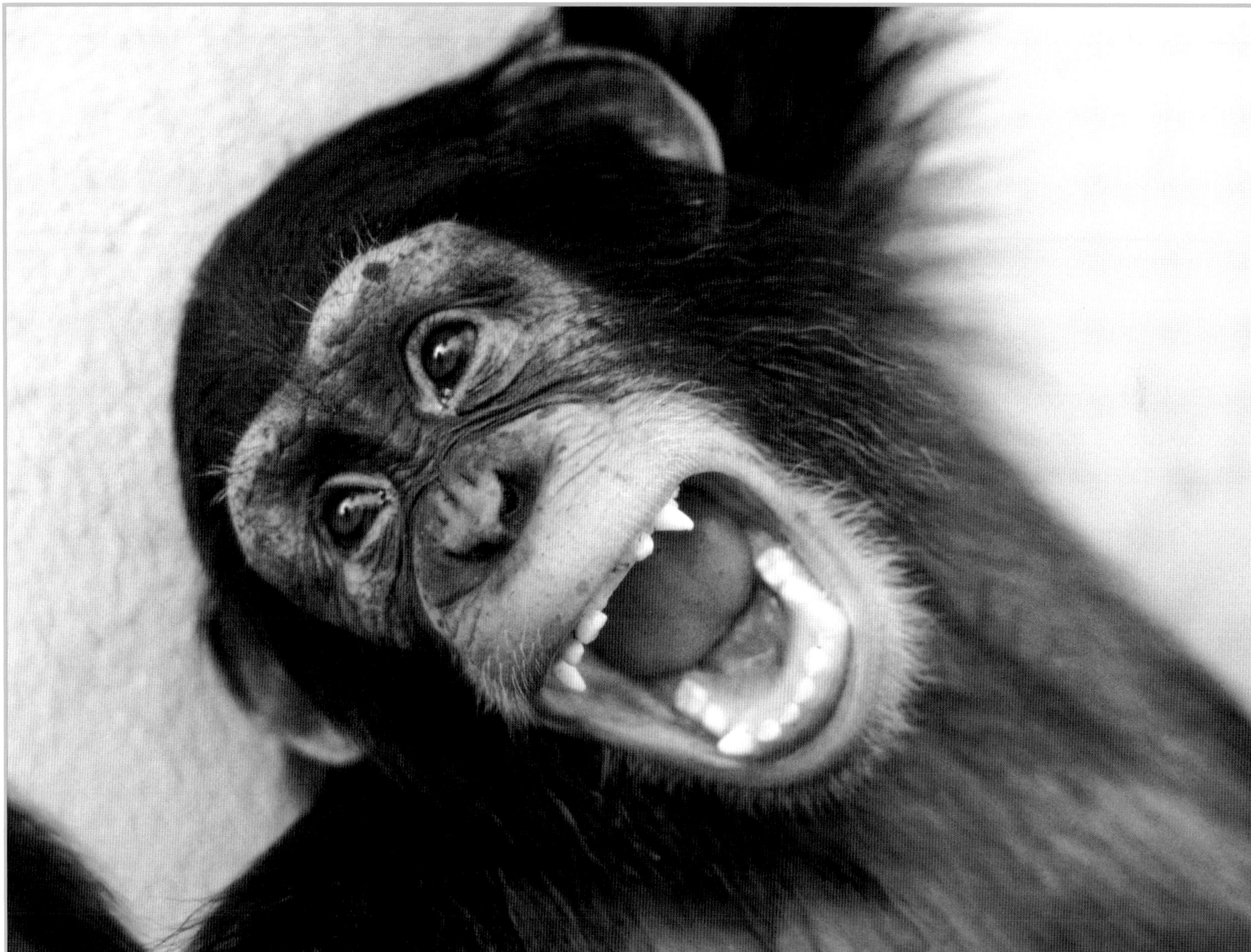

Funny faces

Dutifully, I kept records of every facial expression observed during my study of the bonobos at the San Diego Zoo. My intention was to make a complete list of the species' expressions. Like all primates, including humans, bonobos have facial expressions that signal emotions, such as anger and sadness. Soon it became obvious to me, however, that the juveniles of this species had an entire set of extra "faces" without any connection to emotions or communication. They would make strange grimaces without even looking at each other, aiming them at no one in particular, changing them all the time. I began to call these random expressions *funny faces*. It was a game to them, but a very instructive one since it showed that bonobos have remarkable voluntary control over their facial musculature. Like human actors, they can play with their faces, blowing up their cheeks or sucking them in, hooking a finger into the corner of the mouth, or baring just the lower teeth while rapidly moving the jaw up and down. To move the face deliberately, without emotional involvement, is a rare activity in the animal kingdom—it may be one more ability shared between humans and apes.

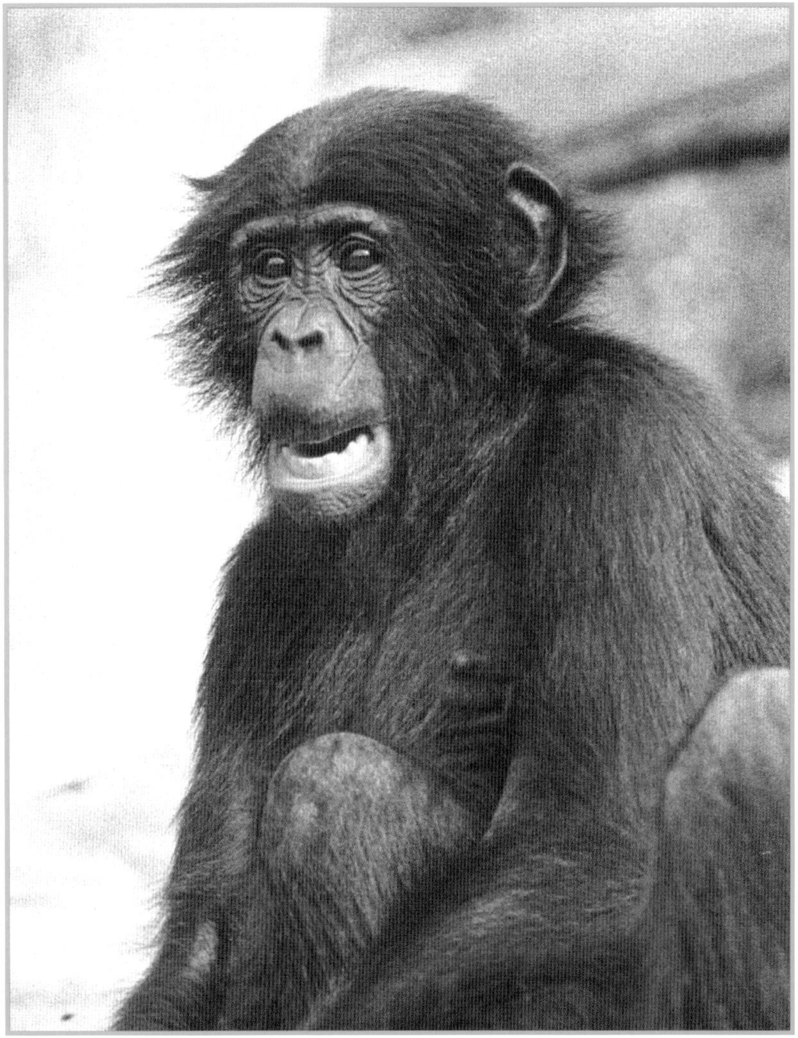

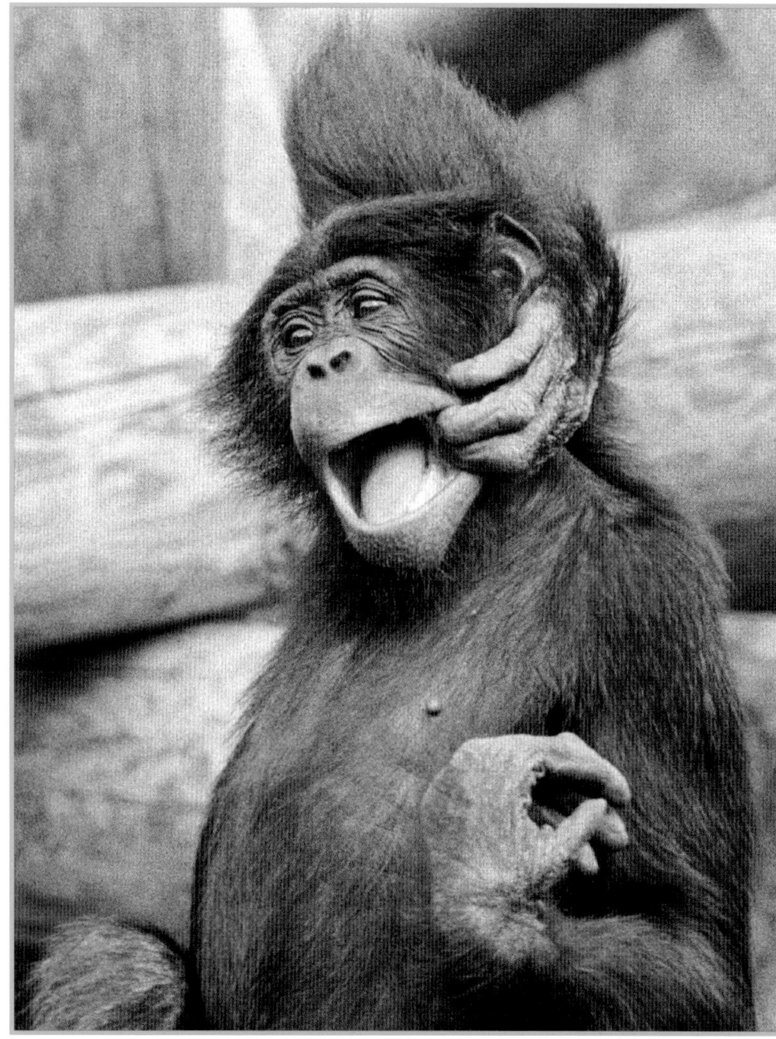

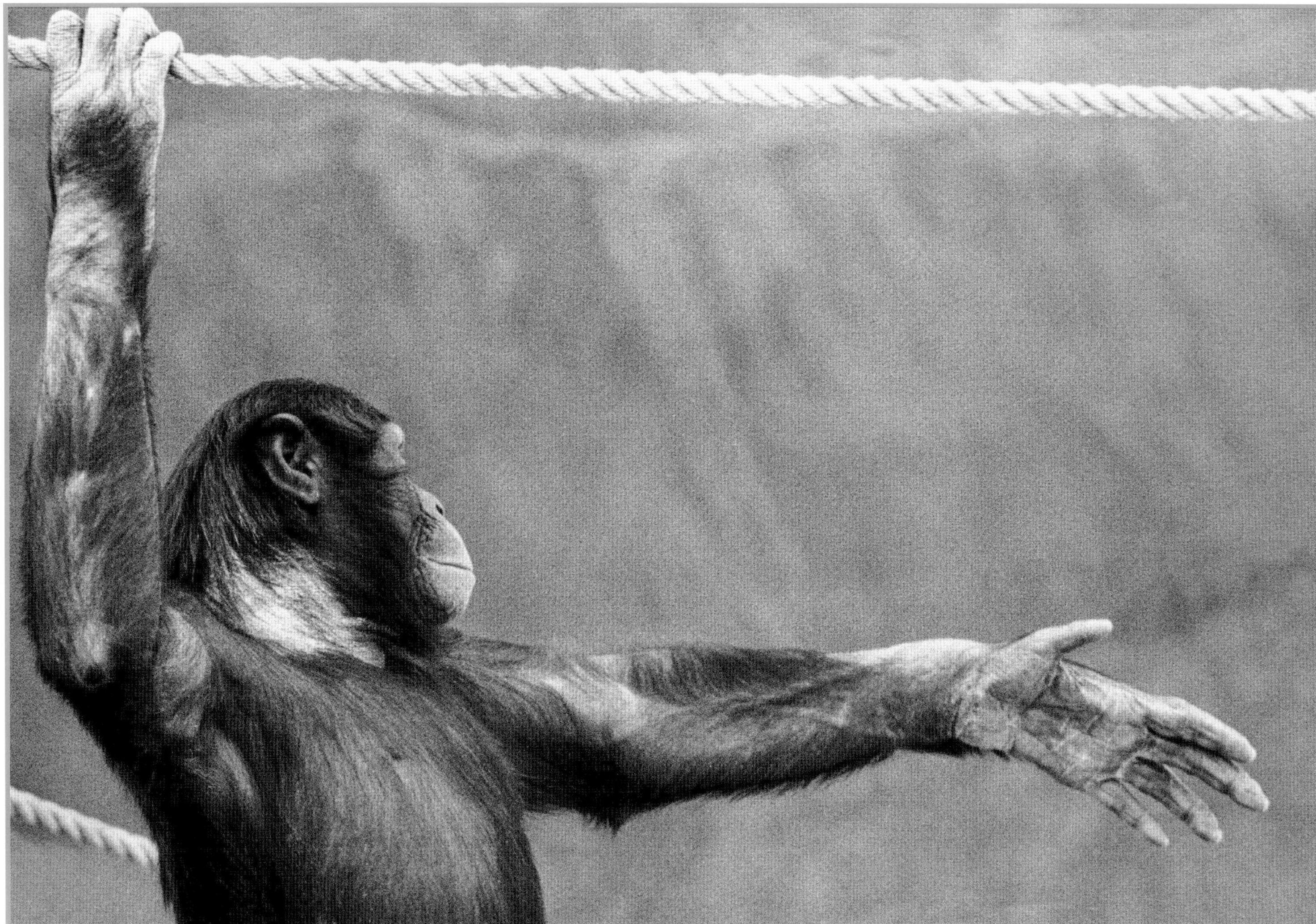

Reaching out

The apes' begging gesture probably derives from cupping a hand under the chin of another ape in possession of food (page 101). Begging for food can also be done from a distance, however, with the palm of the hand turned up. This is the universal supplication by both human beggars in the street and apes interested in someone else's food. However, not all begging relates to food. An ape may beg for support in a fight by reaching out to an ally, or for physical contact, such as nursing or sex, that requires cooperation. In such cases, the hand gesture is not always made with the palm up: here an adolescent bonobo male is trying to get a female to approach him for a mating. The meaning of his gesture was obvious given that he was showing an erection at the same time.

The human heritage of gesticulation has been related by some scientists to language evolution. Symbolic communication may have started not with grunts and speech but with ever more sophisticated gesturing. Hand gestures lend themselves to voluntary control and—as in sign language for the deaf—permit complex forms of symbolic communication.

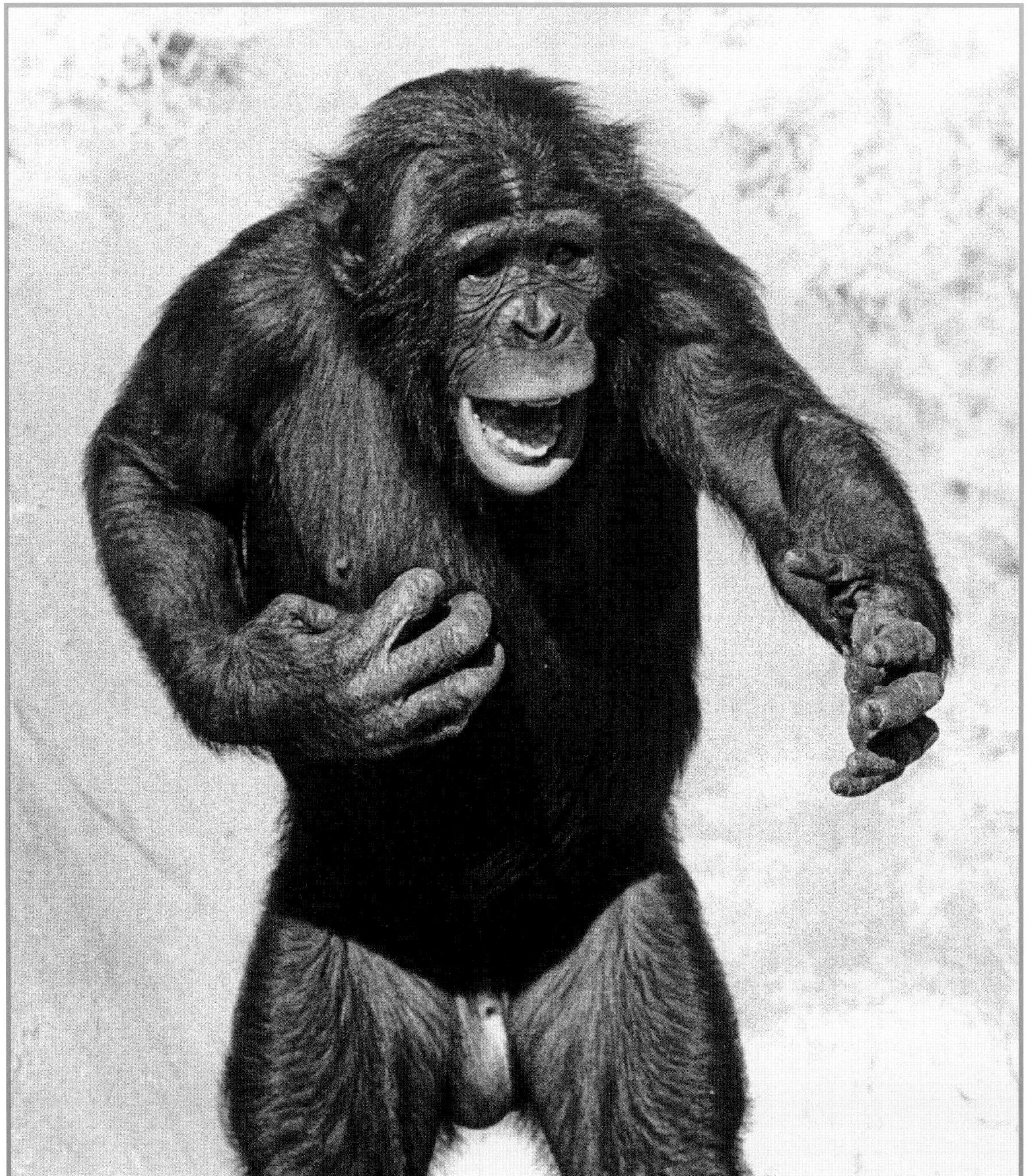

The frown

Puckered eyebrows have been considered uniquely human ever since Charles Darwin, in his admirable book *The Expression of Emotions in Man and Animals* (1872) made the point that the frown is absent in apes. He had tried to aggravate apes with a frustrating task yet failed to get them to frown while concentrating. Only when he tickled a chimpanzee's nose with a piece of straw did Darwin obtain a few vertical furrows between the eyebrows.

Chimpanzees do frown, however, and bonobos have an even stronger contraction of the corrugators, the muscles that bring the eyebrows together. The male bonobo in the large photo is hooting excitedly to draw the attention of caretakers who are bringing sugarcane. With his right hand he caresses his own nipple (which is this male's way of reassuring himself); with his left hand he begs for the food, and his eyes are pulled together and angled in a frown. The small photo shows the face of a juvenile bonobo who is charging two others. While the other two stand united with arms wrapped around each other in defense, the aggressor's face lips are pulled tense with anger, and he shows a frown that gives his eyes a piercing expression.

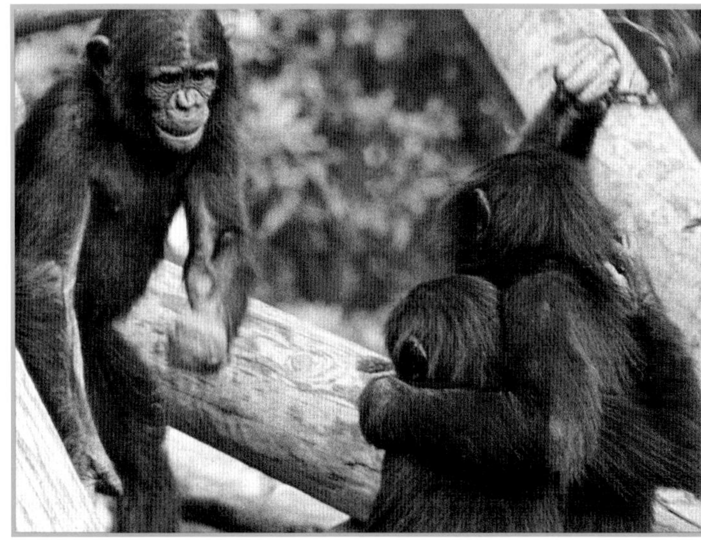

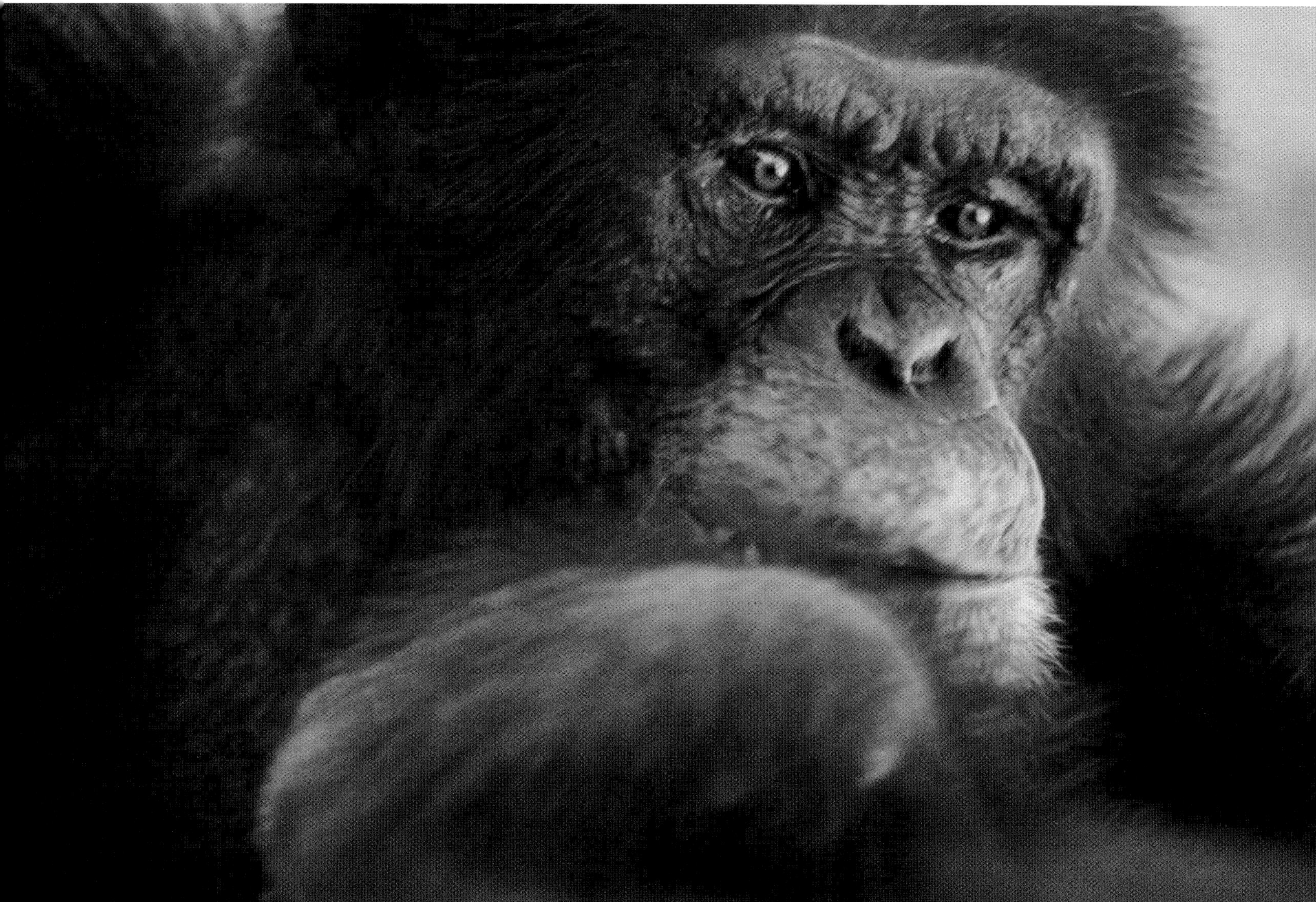

Eyes full of worry

Socko's eyes often had a worried expression during the period in which he and another male, Bjorn, vied for the top spot in the chimpanzee group at the Yerkes Primate Center Field Station. Bjorn was a smart and tough opponent: hence things were not going well for Socko, who regularly suffered gashes inflicted by Bjorn's canine teeth.

Eyes are the most expressive part of the ape's face. As in humans, they provide a "window into the soul." Some people might object to such an observation as anthropomorphic, yet in relation to an animal as similar to us as the chimpanzee, and given the evolutionary closeness between humans and apes, continuity seems in fact a safer assumption than difference. Socko did not always look like this: when his social status improved, the tense stares disappeared. For this, he had to wait until a third male grew up, one who was willing to take his side against Bjorn.

Ancestral traits

Bonobos and chimpanzees are so similar that we might call them sibling species. Not so long ago bonobos were still known as pygmy chimpanzees, but now they carry their own distinct name. Both apes are very close to us, closer, in fact, than the third African ape, the gorilla. Nevertheless, science is keen on comparisons with the gorilla, since similarities between bonobos and gorillas may point out the retention of primitive traits from a common ancestor.

At first sight, the lanky bonobo seems the opposite of the stocky gorilla, but bonobos and gorillas share a few traits that are absent in the chimpanzee. The rounded heads of juvenile bonobos, for one, make them look like young gorillas. And, like gorilla babies, bonobos are born with a dark face (large photo), whereas chimpanzees are born with a light-colored face that turns dark only later in life (page 34). Most telling, bonobos have a gorilla-like nose with thick-walled nostrils, as shown in the small photograph, a feature absent in the chimpanzee.

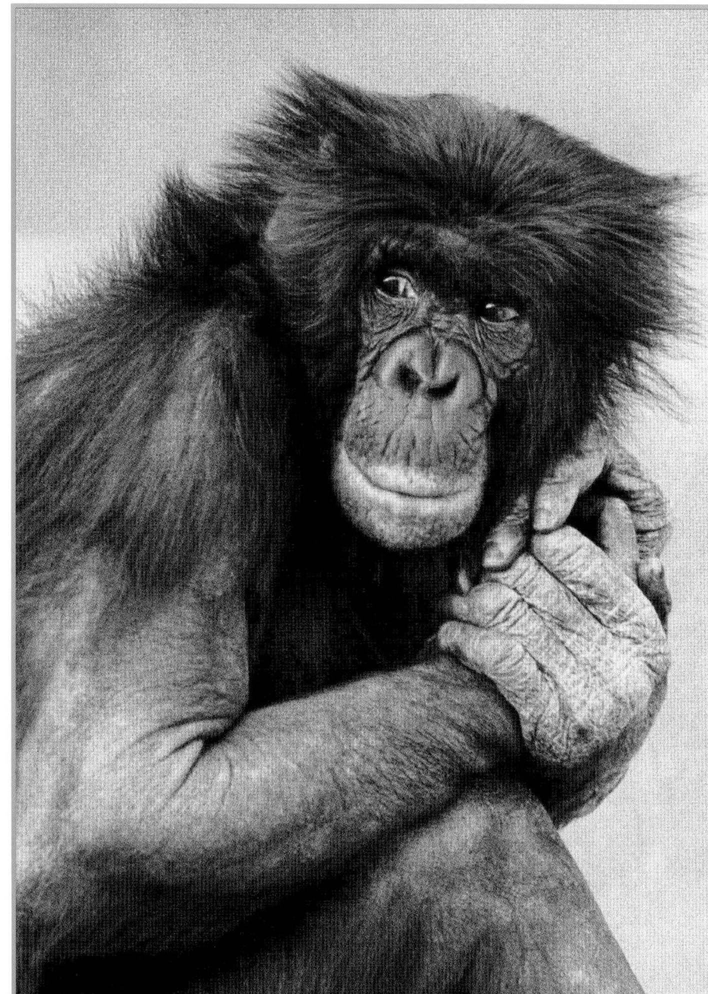

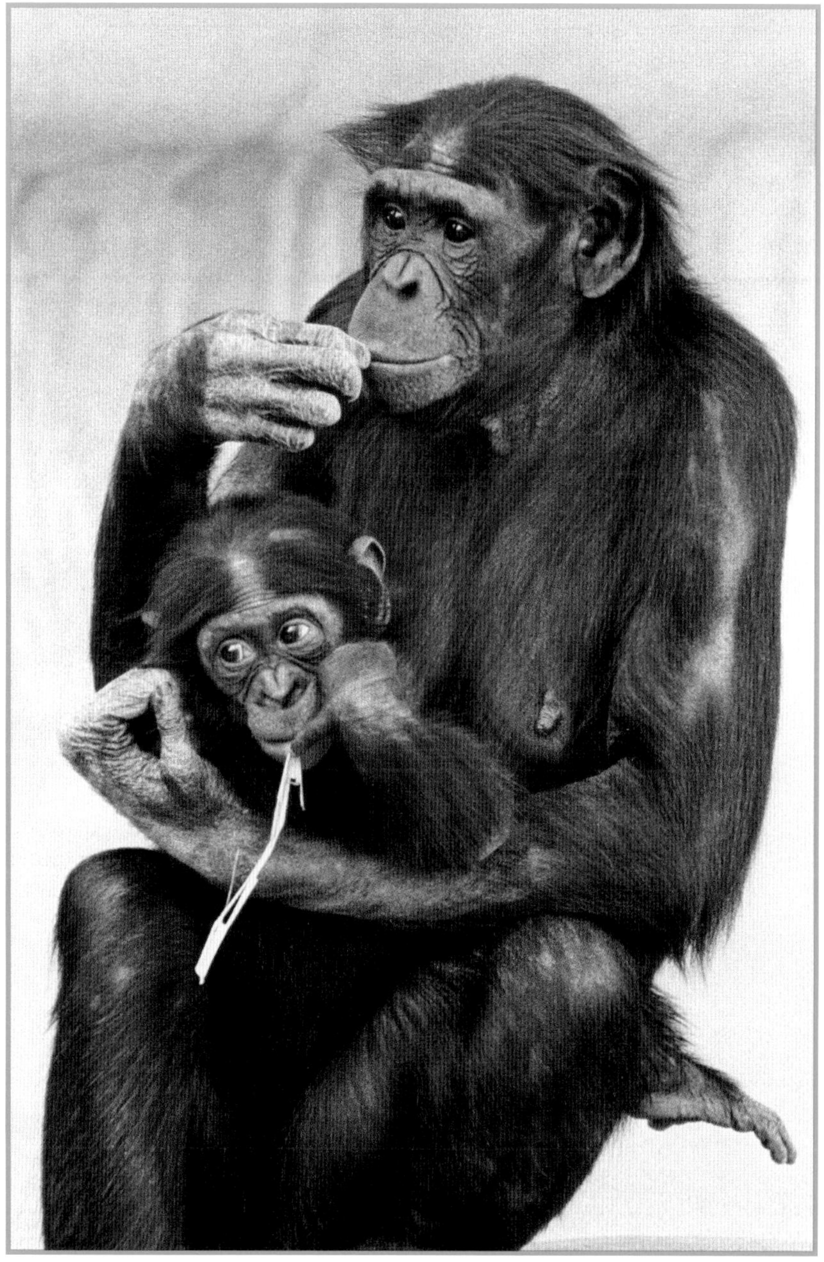

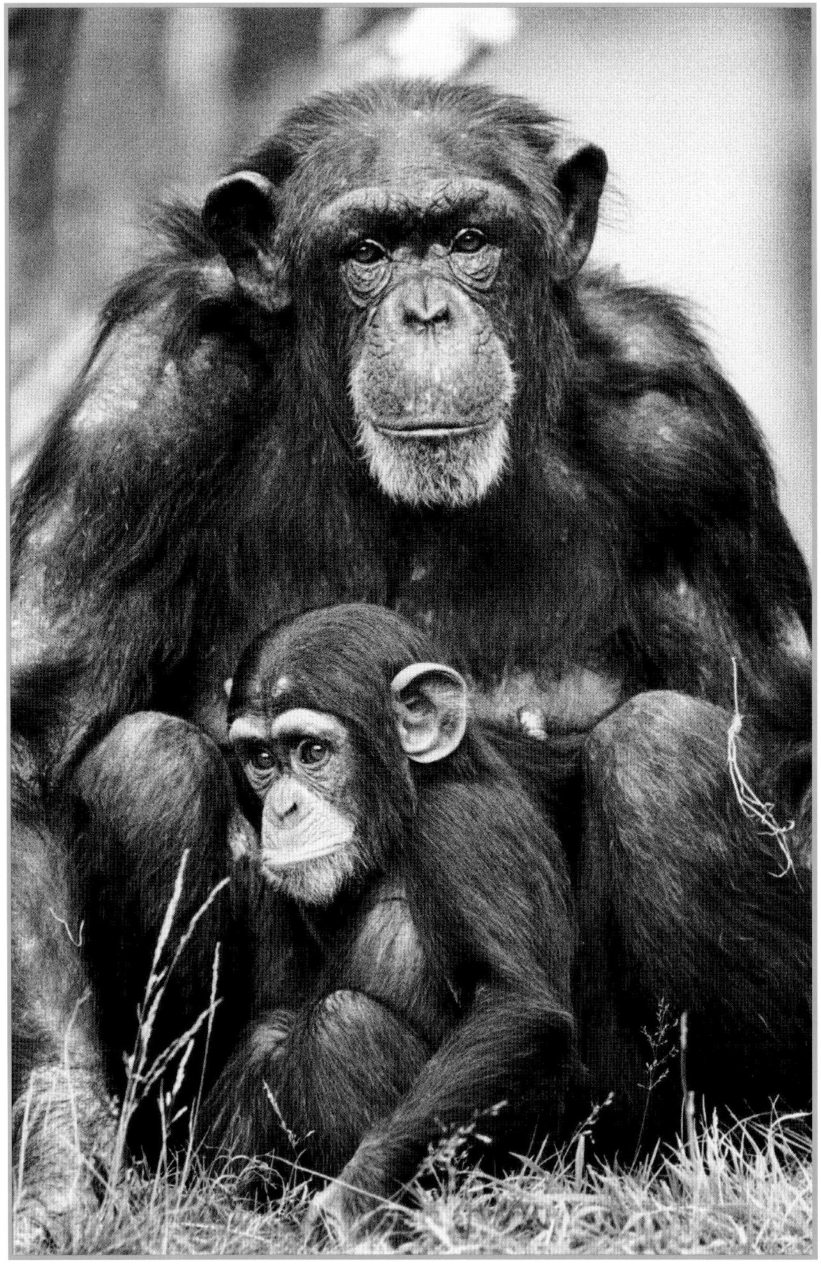

The matriarch

I named her "Mama" because of her matriarchal position in the Arnhem Zoo chimpanzee colony. There was a time when all females obeyed her orders and all males saw her as the final broker in their political struggles. Mama was able to mobilize the whole community behind a male whom she favored for the top position, and she would punish females who dared to side with a different contender.

Of all chimpanzee personalities I have known, Mama's is undoubtedly the most impressive: she never showed a shred of hesitation and was never afraid to step into a conflict, even though she certainly was physically incapable of defeating grown males. If tensions between two males had escalated to a point at which an actual fight was the only option left, but neither dared to deliver the first blow, both would rush to Mama and sit in her arms, screaming at each other. Fearless self-confidence combined with a maternal attitude put her at the absolute center of the colony's power structure. Her daughter Moniek knew this all too well and took great advantage of having such a respected mother, who would protect her against anyone and anything. Mama is still alive but has aged in the twenty-five years since the first photograph was taken. In the small photograph, which shows her chewing an acorn, she is estimated to be forty-five years old.

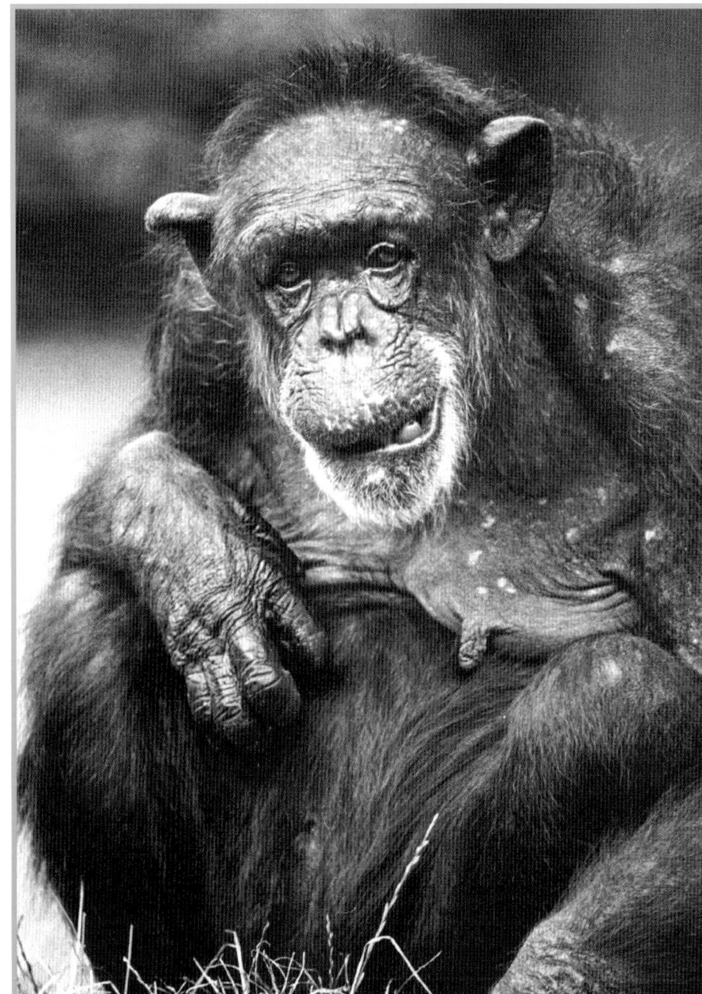

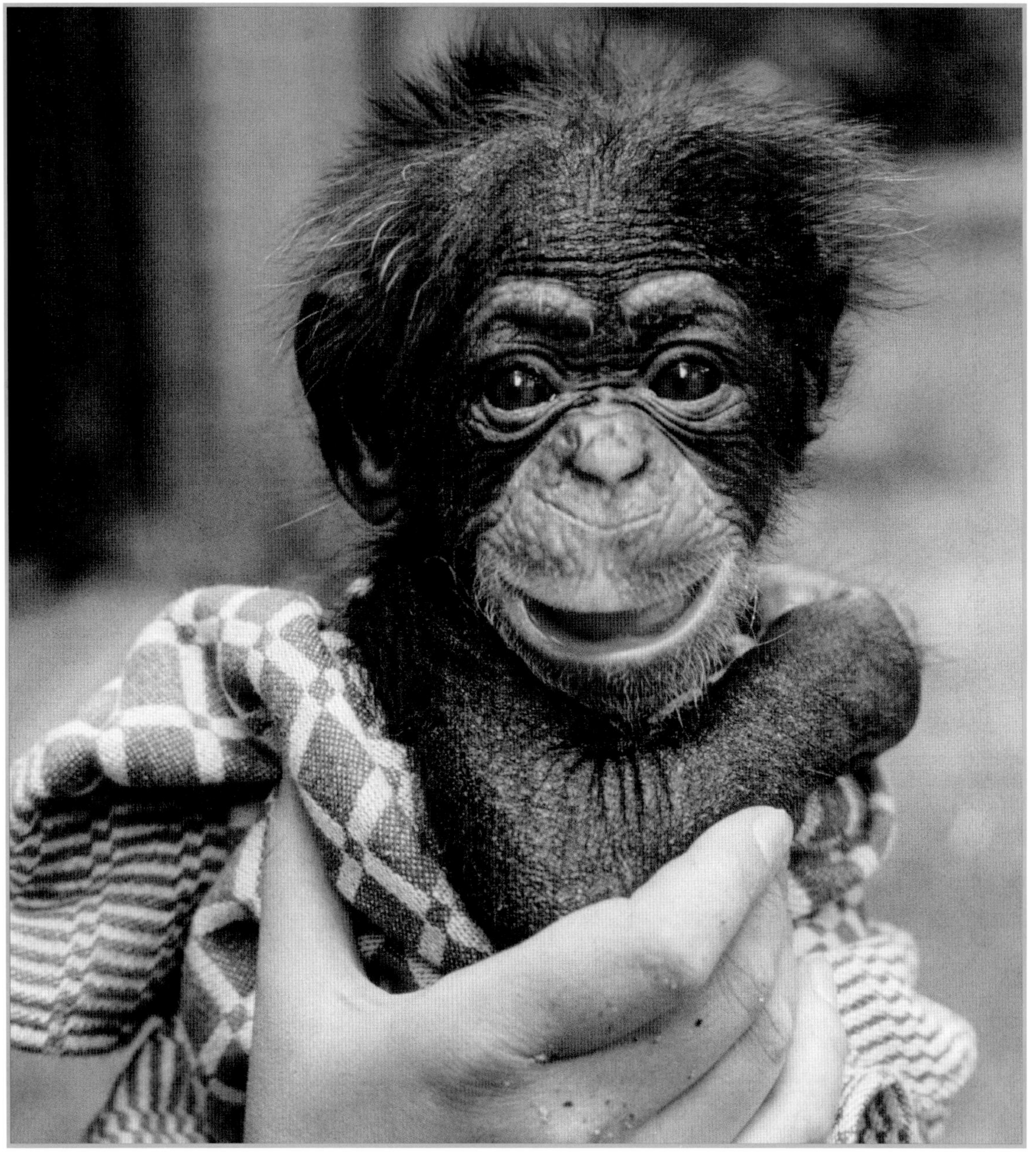

Simply irresistible

Baby apes are almost impossible to see up close, since their mothers try not to expose them. The mother keeps the baby tucked safely under her belly, and whenever someone tries to sneak a peek—and certainly if one aims a camera at the newborn—she will turn her back. Shown in the small photo, a rare glimpse of a sleeping infant is given by a sunbathing female in the sand directly under our observation window (the mother had not heard me opening it).

Baby apes in human care are a different matter. Before Roosje was adopted by a female of her own species (pages 142–43), we fed her with a bottle. This allowed us to interact with her and let her get used to people. She didn't mind being handled and photographed; here, in the large photo, she looks straight into the camera while the caretaker holds her up. I have always found infant apes to be totally appealing. Maybe it is their hair or the entirely black eyes (humans are the only primates with white around the iris). They have the same rounded features and large eyes as infants of almost any species, a phenomenon that made early ethologists speak of a *Kindchenschema*, or "infant appeal." Such special physical features of the newborn trigger protective and caring responses, an effect so universal that it also works between species.

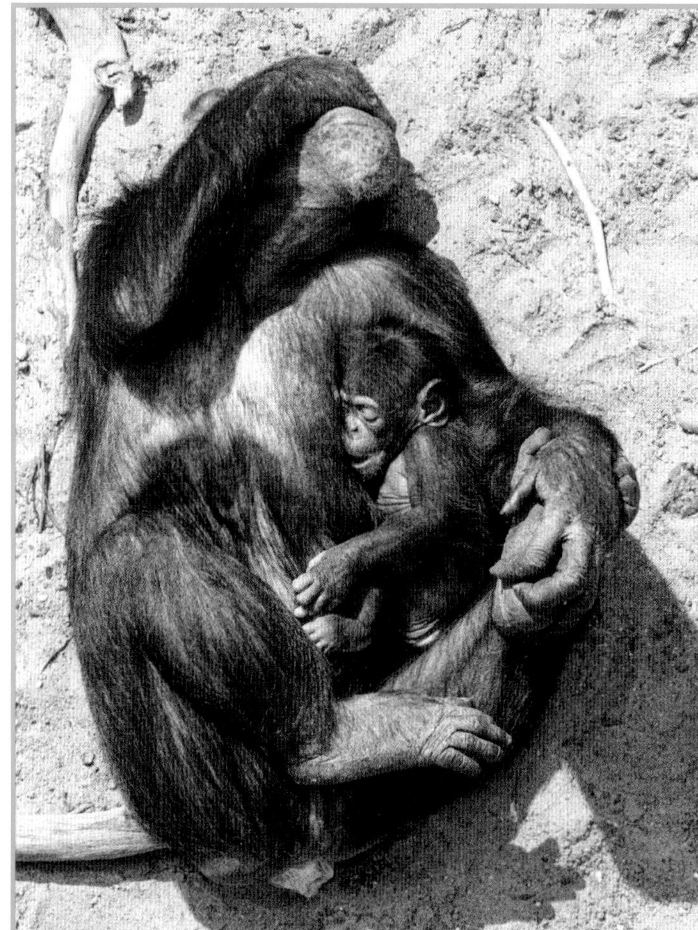

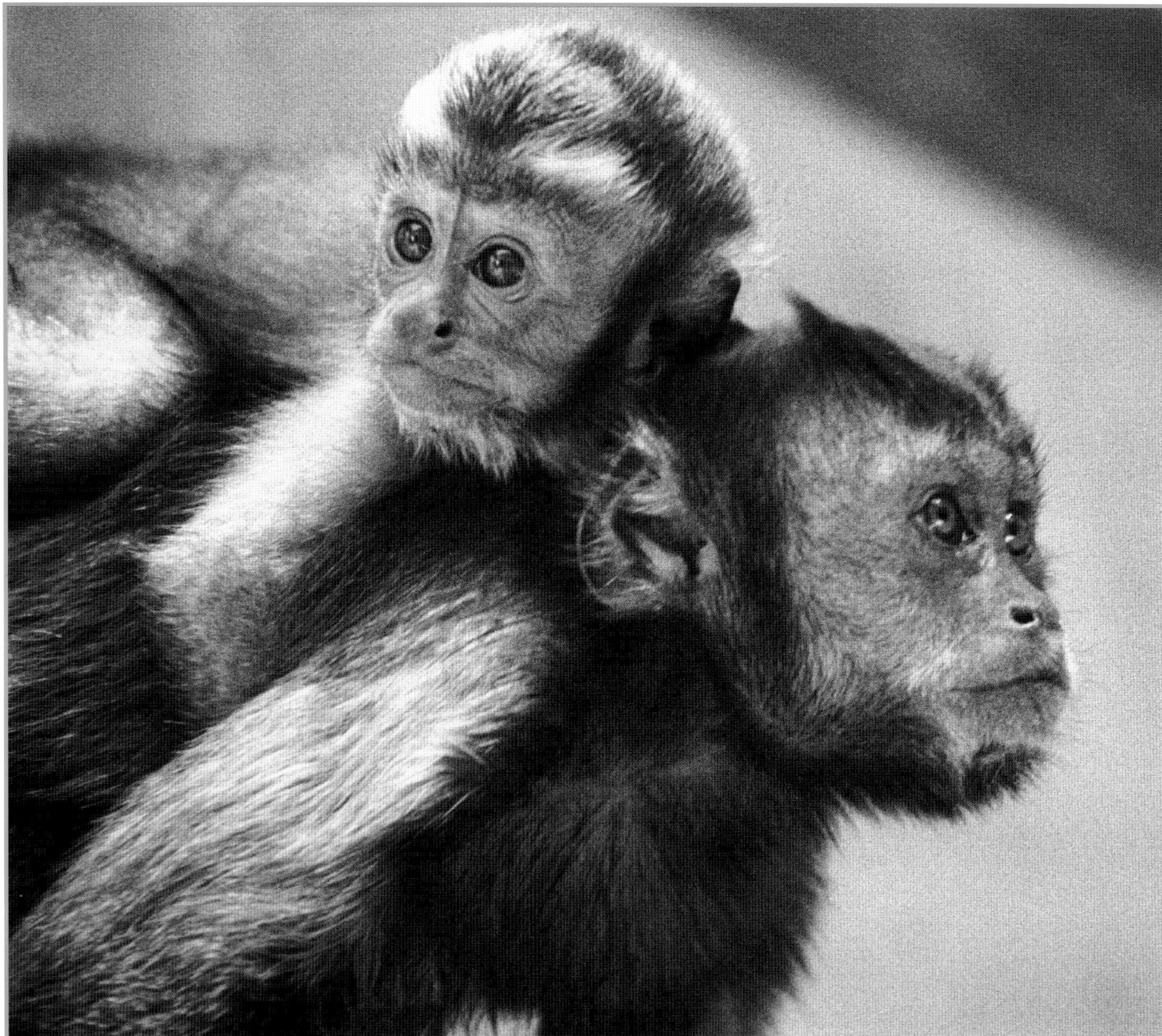

Securely attached

The tufted or brown capuchin monkey delivers a helpless, blind infant after a pregnancy of about 160 days. The infant opens its eyes within five days (small photo). Excellent at clinging to fur with hands and feet, the baby is carried on its mother's back from the first day, occasionally nursing under her armpits. Mom endures dramatic temper tantrums during the weaning period, when her infant is about one year old. In our studies on capuchins, we found that as in humans, one can divide the bonds between mothers and offspring into secure and insecure attachments. The former are characterized by more affectionate behavior and less conflict between mother and offspring, whereas the latter show the reverse. Securely attached monkeys develop a calmer, less agitated way of handling social situations when they grow up.

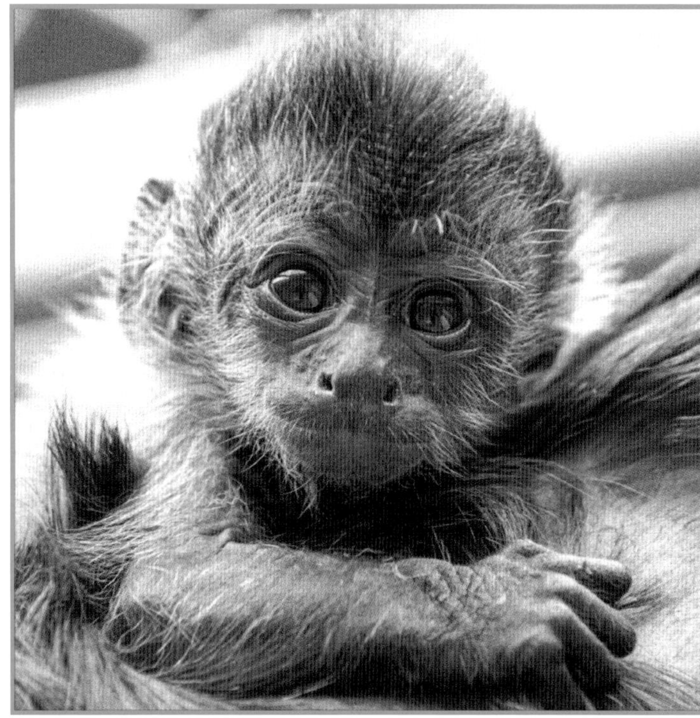

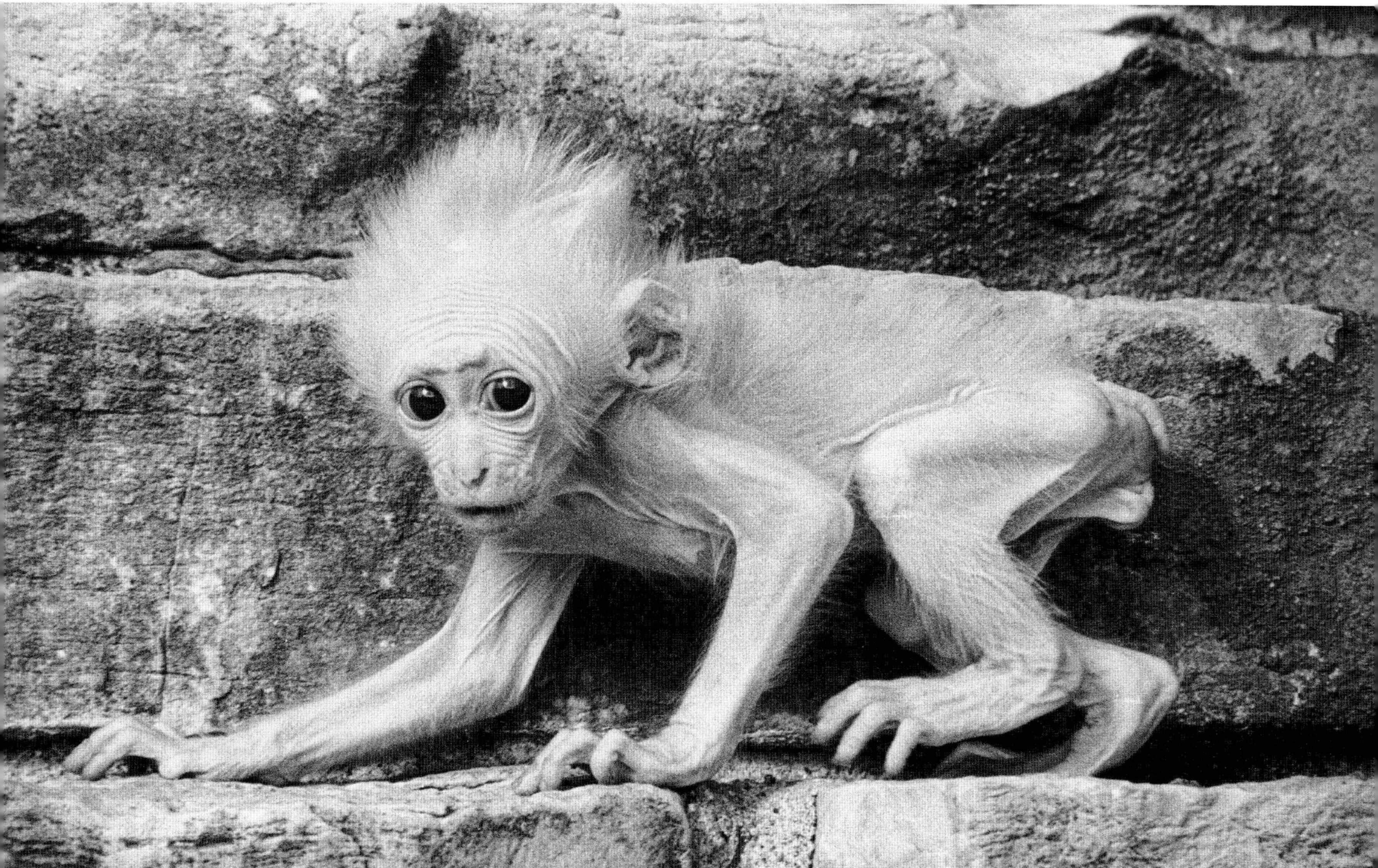

E.T.

Stump-tails are born with a white coat that makes them stand out among the much darker adults. We call them "E.T.s," after the movie character, because of their out-of-proportion head atop a thin neck and their alien appearance. Only after about six months does their coat begin to turn red-brown or black. They also seem more dependent and fragile than most primate infants. The male in the small photo is already three months old, clambering around unsteadily on long limbs with his genitals dragging along the ground. As if this is not enough, he draws attention to himself by squealing at the top of his lungs. And attention he gets! All eyes of the group are glued to any white infant that is walking loose, even though no one will approach it, let alone touch it—perhaps out of respect for the mother, who of course shows no hesitation picking up her pride and joy. The inhibition of others against touching newborns is so extreme that the infant temporarily seems to rule the group: even the highest-ranking male or female will retreat for an approaching E.T. I have seen newborns hobble up to a sleeping group of a dozen monkeys and scatter them all!

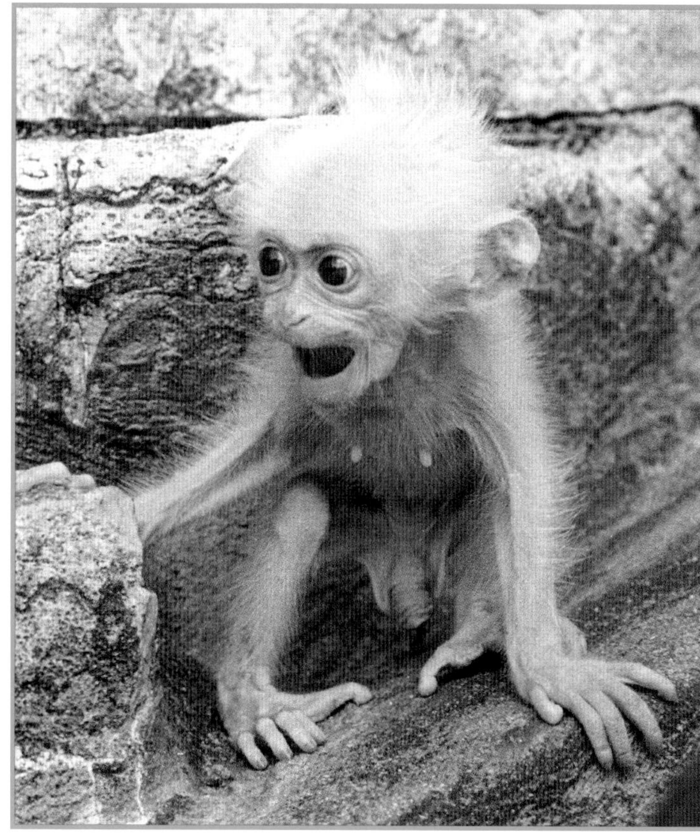

New life

According to a theory known as "attention structure," attention in a primate group goes first of all to the top individual, then to the next in rank, and so on. Being dominant is thus equated with being interesting and important. If one actually measures attention, though, one finds that infants win hands down over those magnificent alpha males. Especially to female primates, there is nothing more fascinating than a baby. The newborn stump-tailed macaque shown here, moreover, stands out thanks to its whiteness. Everyone follows his first movements and wants to sit close to the mother (the central female with the prominent nipples) to inspect him from all angles. The surrounding group softly coughs and grunts each time the baby does something surprising (such as sticking his foot in his mouth), as if commenting on the marvel of new life.

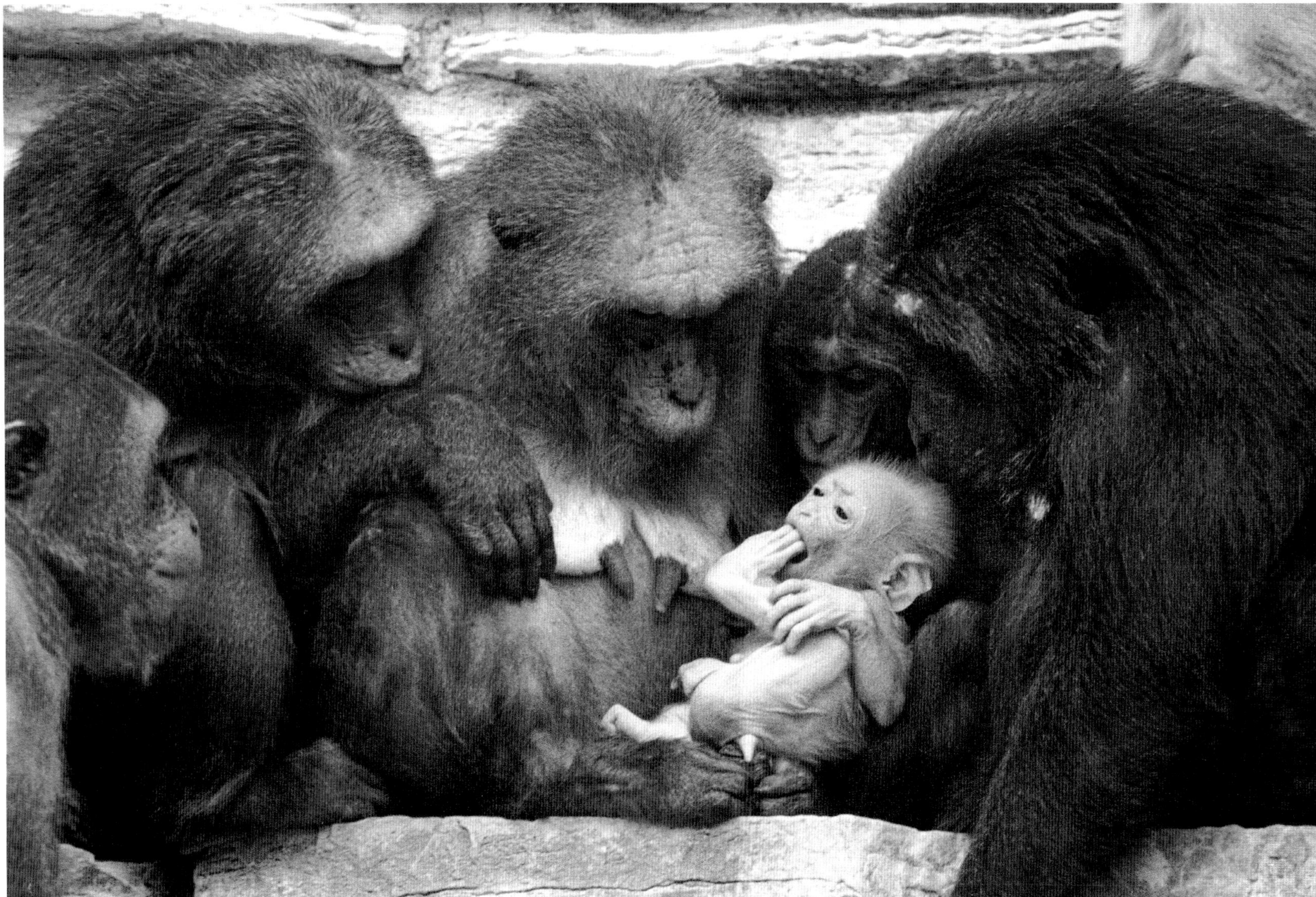

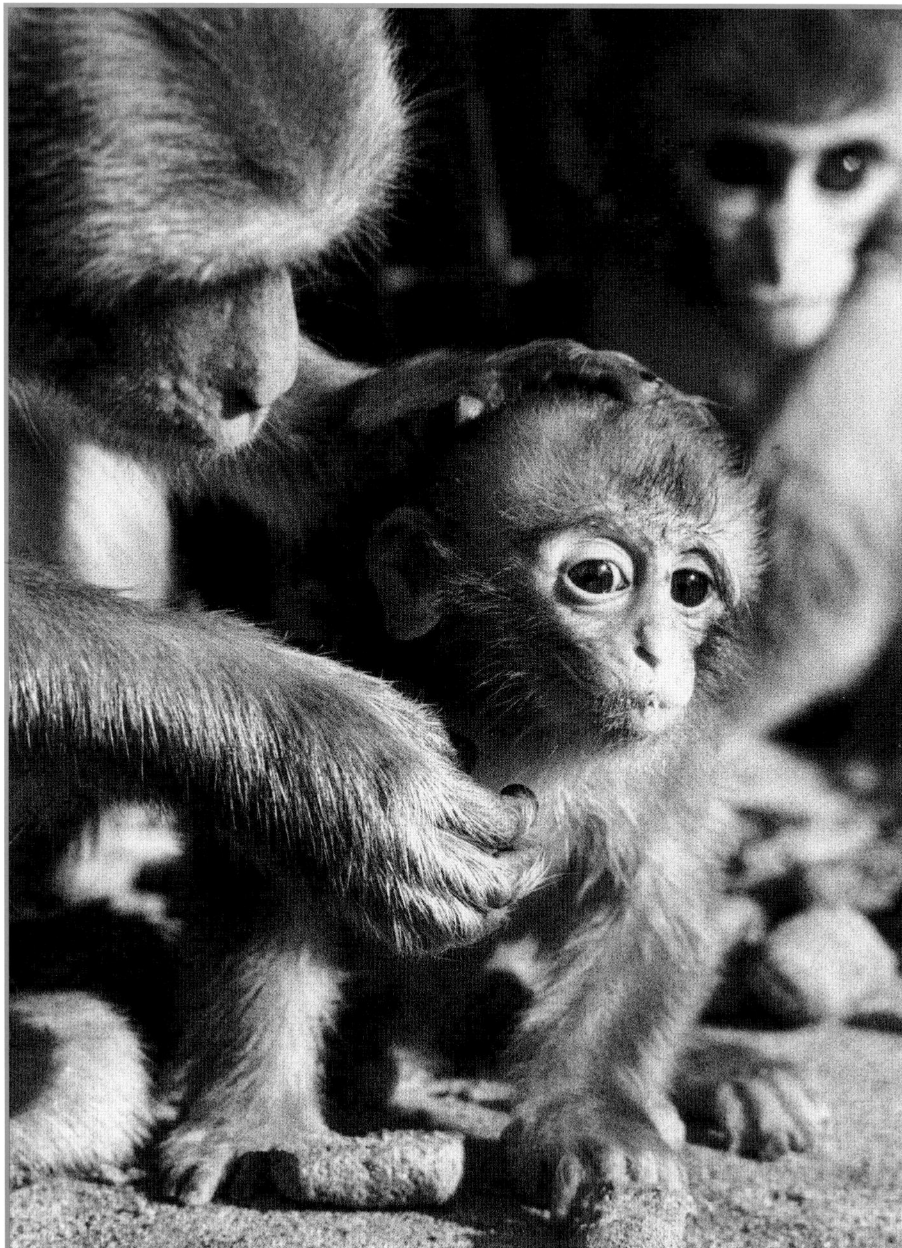

Azalea

It took us several weeks to notice there was something wrong with Azalea, a descendant of a dominant rhesus matriline whose members all had names starting with *A*. Azalea was slow and clumsy and kept close to her mother, Ape, who was extraordinarily patient and protective. In the large photo, Azalea is three months old and clearly doesn't look normal. It was only much later that we analyzed her chromosomes and learned that she was trisomic. With an extra chromosome 18, her condition was not unlike that of human Down's Syndrome, which is due to trisomy of chromosome 21. In another parallel between humans and primates, Ape was aging when she gave birth to Azalea.

Azalea was mentally retarded and socially naive, and she lacked motor coordination. Unable to chew hard food, she was fussed over by her older sister, Apple, who carried her around and let her pilfer the crumbs of hard biscuits that she chewed. In the small photo, Azalea (with Apple on the right) is fifteen months old and doing well. By then, her facial features had normalized, and the group had come to accept this unusual but harmless character. When she was close to three years old, however, her condition suddenly deteriorated, and we had to put her to sleep.

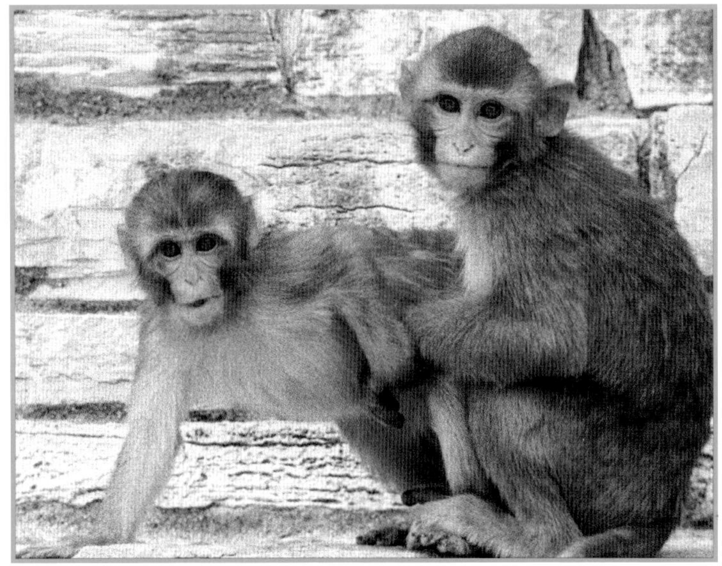

Two in one grasp

Ropey, a rhesus female, developed the habit of holding her own infant together with peers of the same age. In the group among which she lived, other mothers did the same but none as often as Ropey, who showed this behavior year after year with all of her babies. We began to call it the "double-hold." The "kidnapped" infant would usually try to struggle free, looking around for its own mother: within a few minutes the double-hold session would be over. Here Ropey holds her own infant on the right, while the other infant looks away, resisting her tight embrace.

Knowing how Ropey picked out infants for the double-hold may help us understand her behavior. Nine out of ten times, the second infant was from a family that ranked above Ropey's. Since Ropey ranked near the top, there were not many high-ranked kids around, and she had to seek them out. We speculated that she was trying to promote ties between her own offspring and future friends from the upper classes. In the hierarchical society of rhesus monkeys, such contacts may be advantageous.

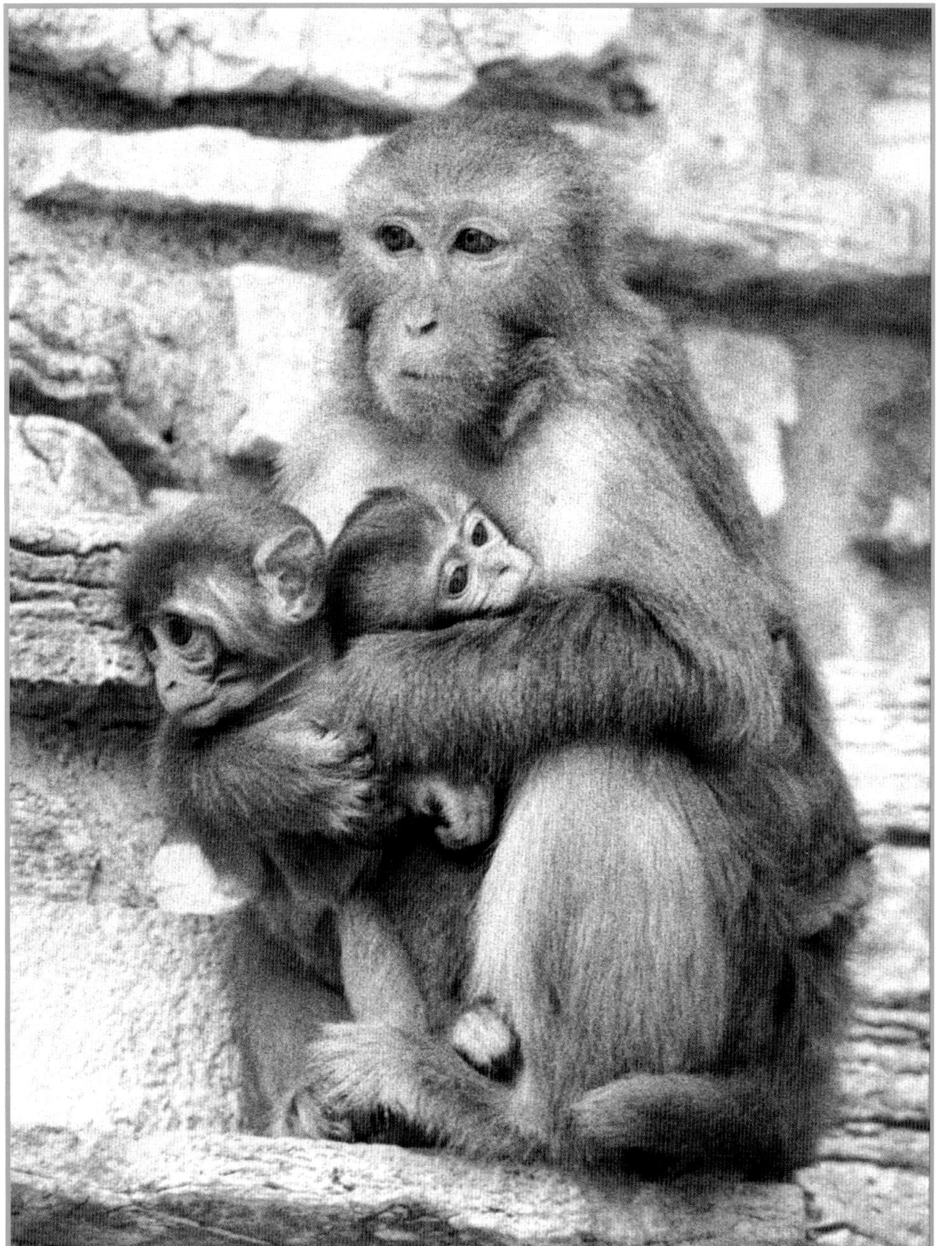

Weaning compromise

Weaning is a veritable battle between mother and offspring: it disturbs a previously secure, peaceful relationship. In chimpanzees this battle starts late, after about three years. The youngster, who until this time has been allowed to nurse whenever he wants, suddenly faces maternal restrictions. If the mom has no time, the child has to wait, yelping and begging, until the mother allows him to nurse. I have heard weaning described by child psychologists as the first human experience of "injustice": from the child's perspective, a God-given right is being taken away. For young primates, things may be the same. They throw violent tantrums, scream and yell, fall out of trees, and roll around on the ground until the mother takes pity on them. In the fourth year, things get worse, because the mother increasingly blocks nursing attempts with actions such as sitting with her arms tightly folded over her chest. For the youngster, it is a losing battle. The weaning conflict winds down in the sort of compromise seen here between a five-year-old male and his mother: he is allowed to suck on any body part except her nipples!

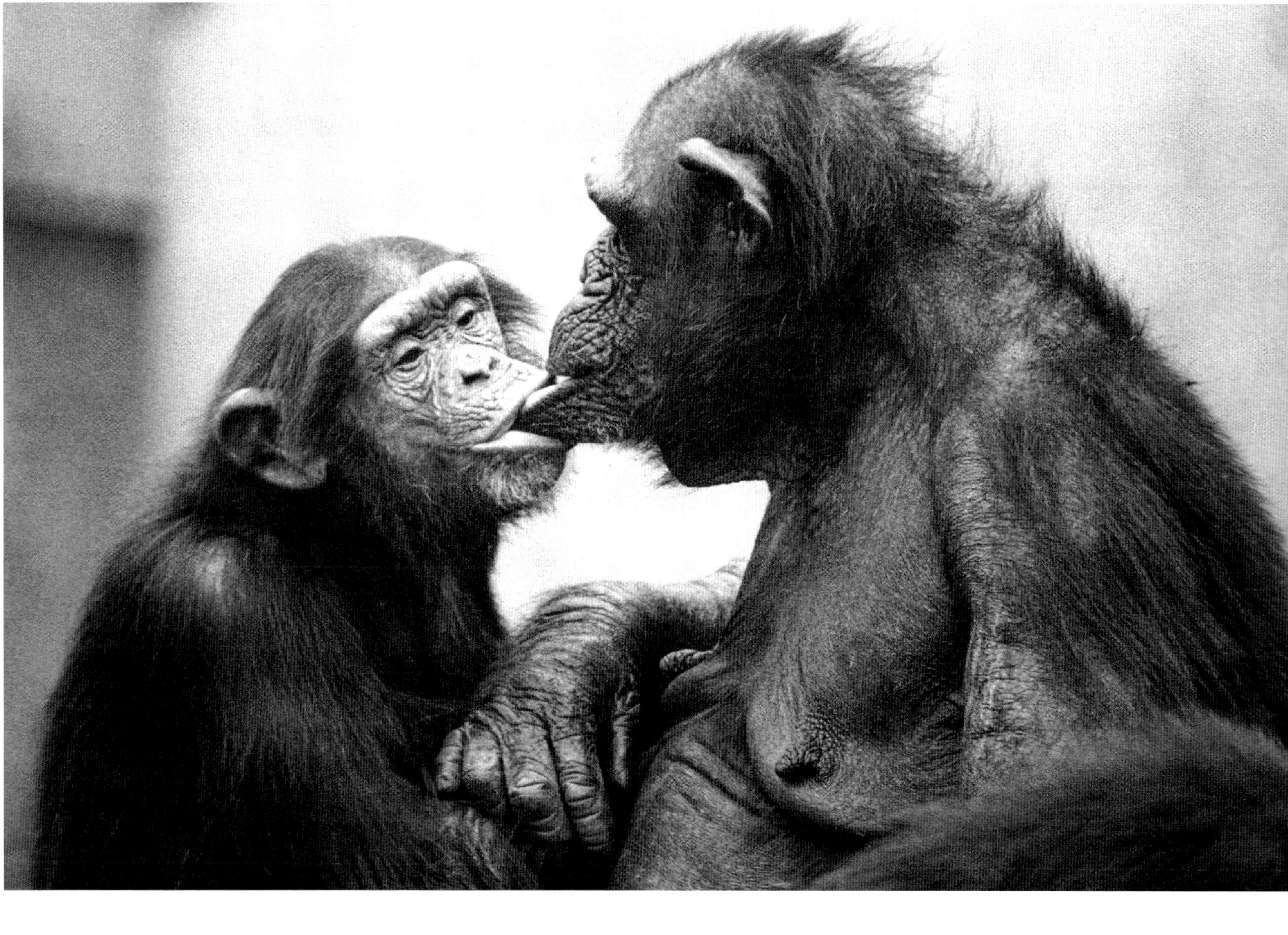

The pygmy chimp is neither

Vernon in a reflective moment at the San Diego Zoo. One can see he is a bonobo rather than a chimpanzee by his black ears, fine hair, light-colored lips, and less prominent eyebrow ridges. Bonobos used to be known as "pygmy chimpanzees," but Vernon is about the same size as males of the smallest subspecies of chimpanzee, hence by no means the dwarf suggested by the old species name. He has a powerful, muscular body and the superhuman strength of all of the great apes, who in hand-to-hand combat have nothing to fear from a human, male or female.

The bonobo was discovered as a species as recently as 1929, making it one of the last large mammal species known to science. The strange name is said to come from a misspelled label on a shipping crate sent to a European zoo, which referred to a village named Bolobo in the Democratic Republic of the Congo, where the species lives.

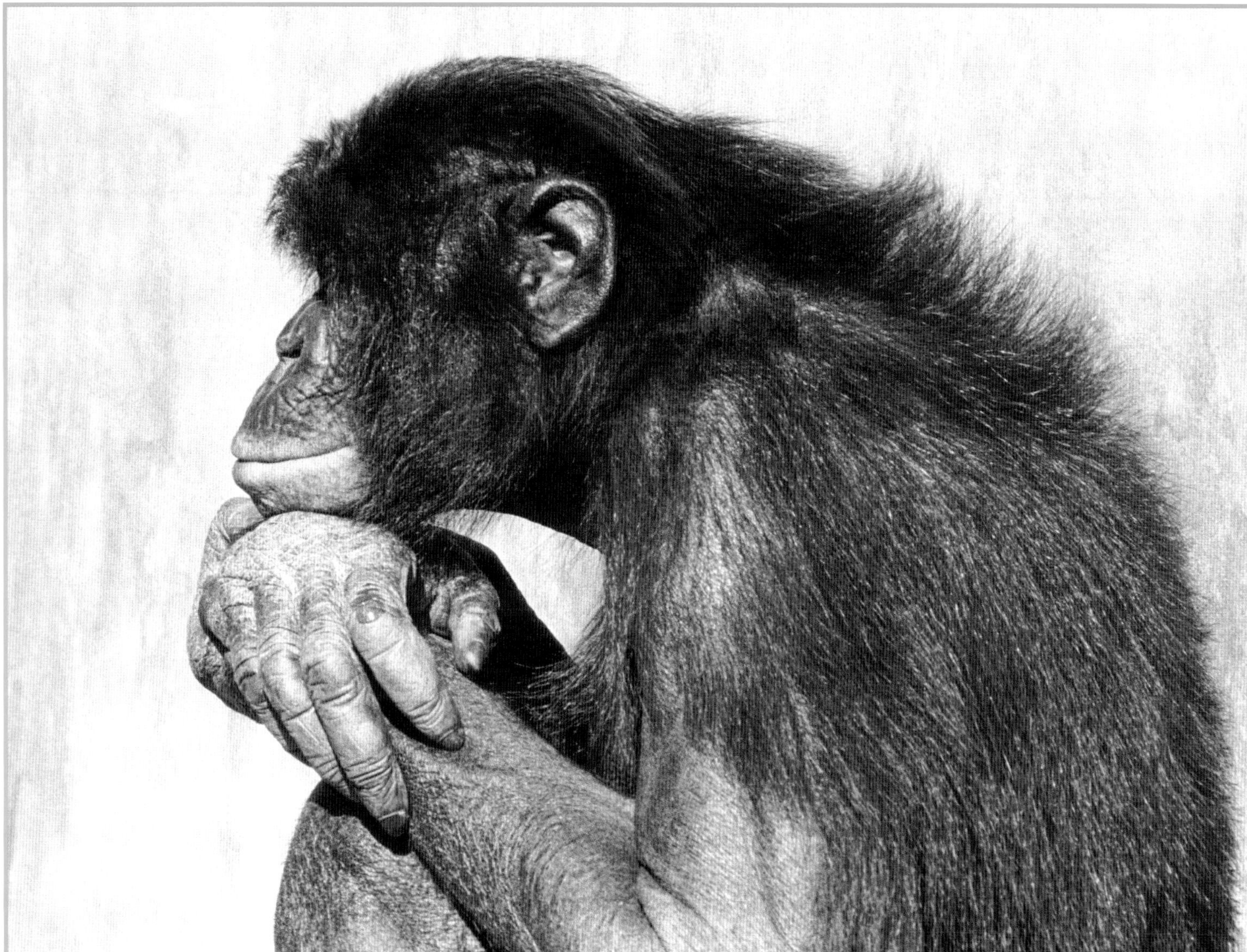

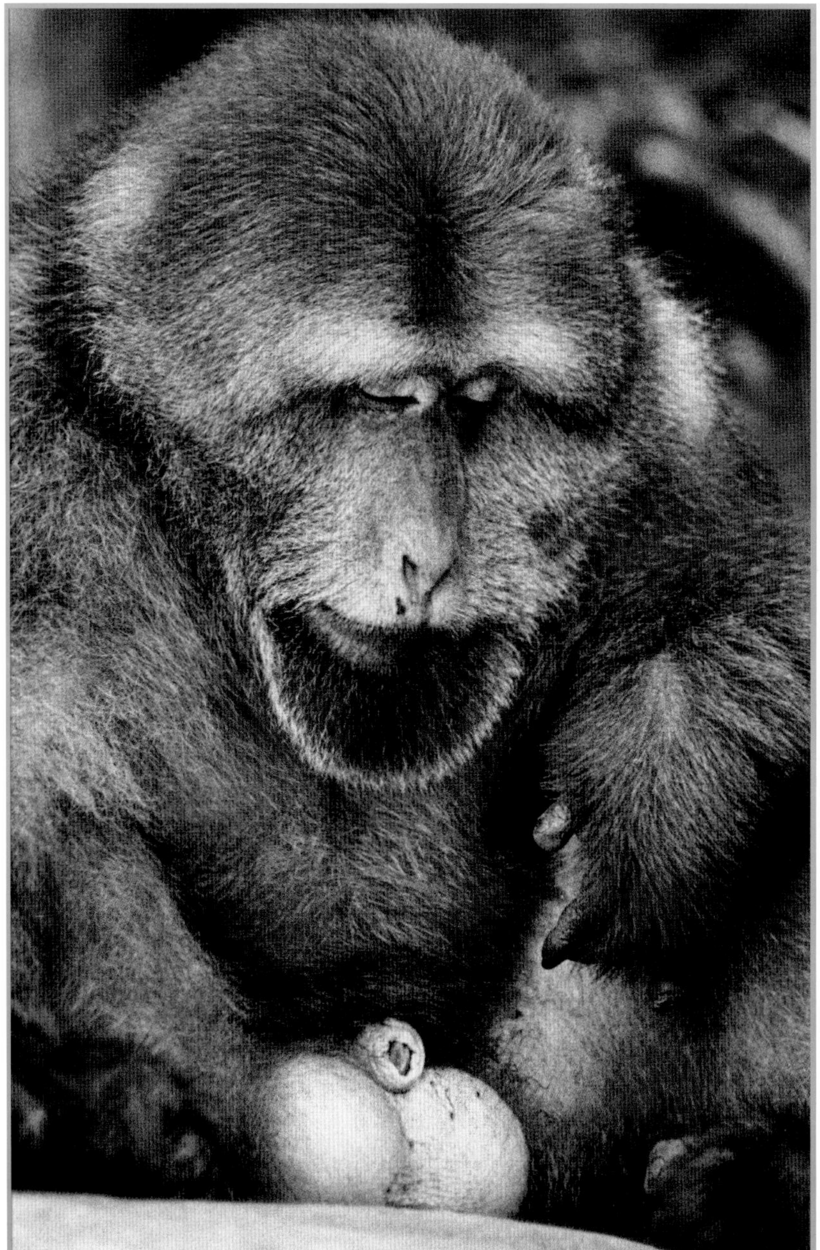

Beneficial balls

A resting Tibetan macaque shows his exceptional genitals: his balls may well be bigger than his brains! Testicle size has become a topic of serious scientific debate. What could possibly be the purpose of testicles capable of producing more sperm than needed for fertilization? One idea is that competition over mates can be replaced by so-called sperm competition. If a female mates with multiple males, the male who produces the most sperm cells will have the highest chance of fertilizing her. Thus, instead of fighting other males and driving them away from fertile females, males may win the evolutionary race by outproducing their rivals in terms of sperm. All that matters in the evolutionary context is how many offspring they sire in their lifetime. As a sign that sperm competition may indeed have taken the edge off sexual rivalries, Tibetan males tolerate each other remarkably well. They get along better than most males in the primate order, occasionally confirming their bonds by mounting each other while squealing loudly (small photo).

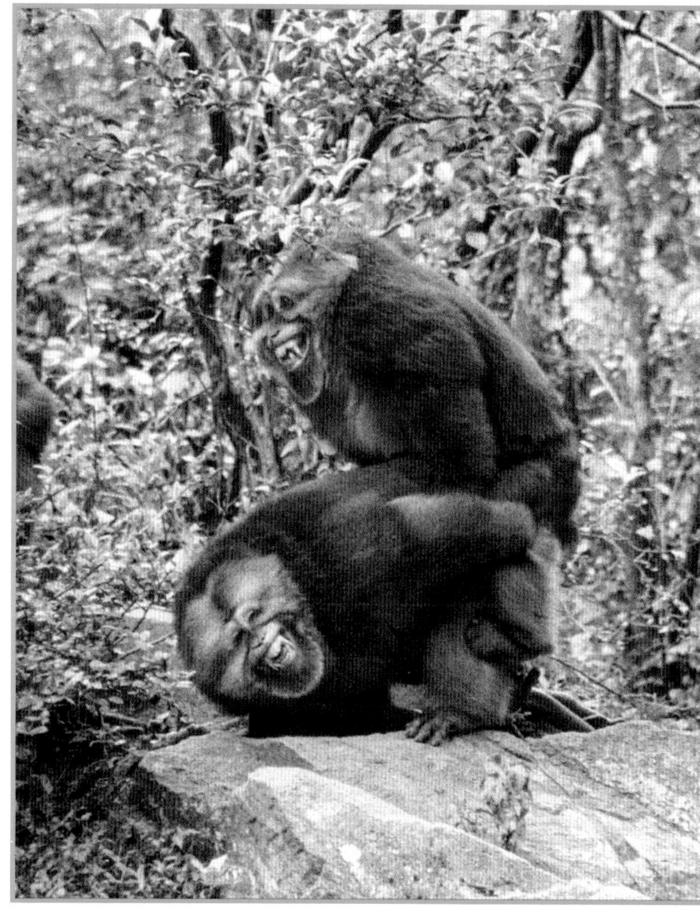

Attractive behind

For the male chimpanzee nothing is more fascinating than a female's behind. Perhaps this obsession is not so different from that of some human males, but in the case of chimpanzees the estrous female sports a balloon-sized pink swelling that humans fortunately do without. I have known male chimpanzees in captivity who wouldn't eat for days on end if one of the females in their group was swollen, which goes to show that for males sex has priority over food.

The female genital swelling consists of water-inflated tissue and signals fertility. If a female in estrus moves in the distance, males will spot her right away and hurry toward her. They then inspect her up close, and if her smell confirms the attractive state she seems to be in, they will invite her for intercourse. If rival males are around, they will engage in complex competitions and grooming negotiations to decide who is allowed to court the female. This is only part of the story, however, because it is then up to her to select which suitor she will mate with.

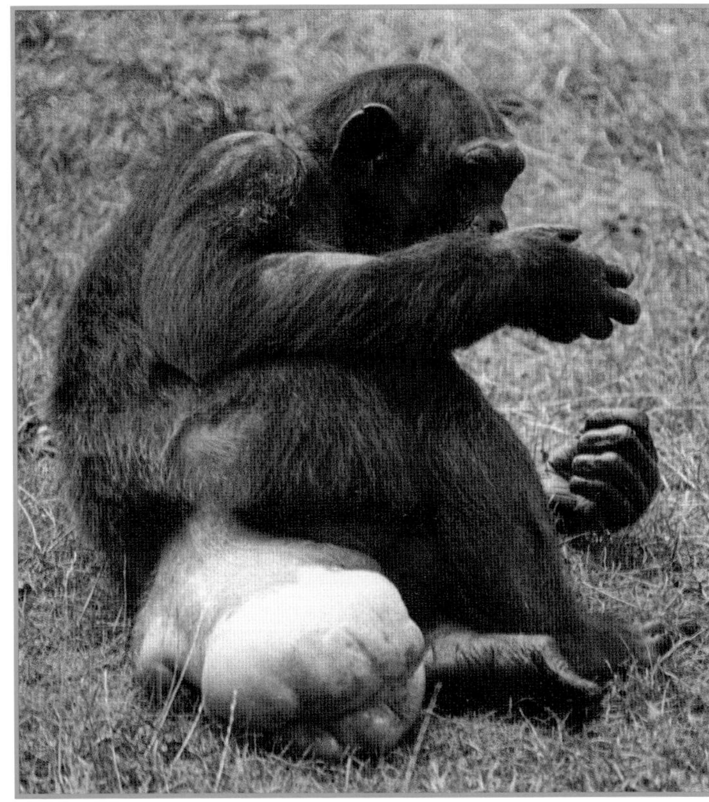

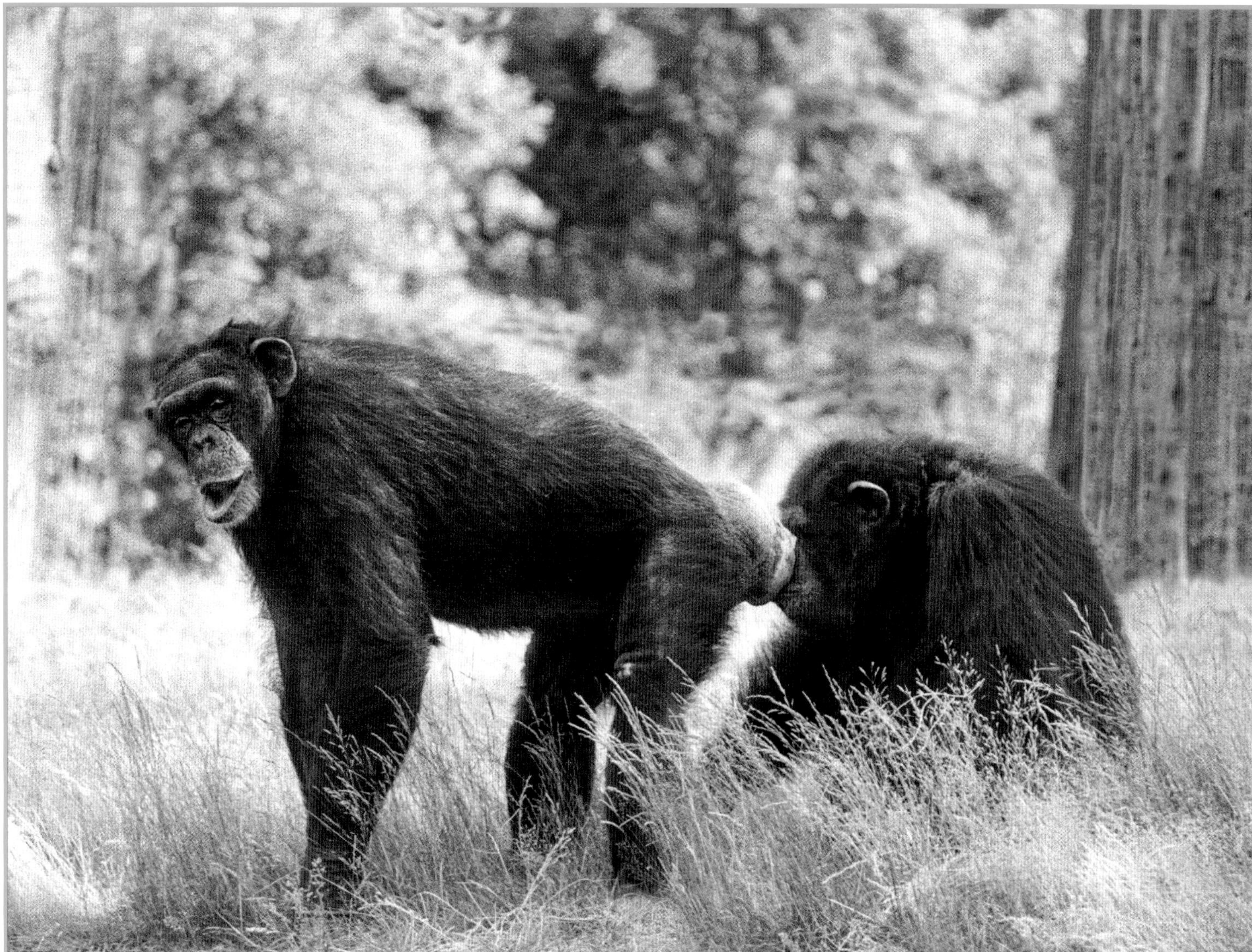

Missionary position

Anthropologists eager to interpret human behavior in cultural terms traditionally have run into problems with sex. The primary function of sex—reproduction through the joining of male and female gametes—is the same for all sexual creatures, and sexual intercourse is fundamentally the same from insects to chickens to humans. Undeterred by this continuity, anthropologists have proposed two reasons why human sexual behavior is special and cultural: we can choose our own copulatory positions, and females of our species experience orgasm.

Both cultural "innovations" exist only in the minds of people, though. Bonobos adopt an incredible variety of positions, including the much-heralded face-to-face position. In addition, bonobo females often squeal and grimace at the climax of sex, leaving little doubt about what they might be feeling. Laboratory studies on female primates engaged in sex offer evidence for uterine contractions and heart rate increases similar to those measured during human orgasm. The female in the present photograph does not show much emotion—she looks rather serene—but then bonobos have many sexual encounters per day, and not every one can be exciting.

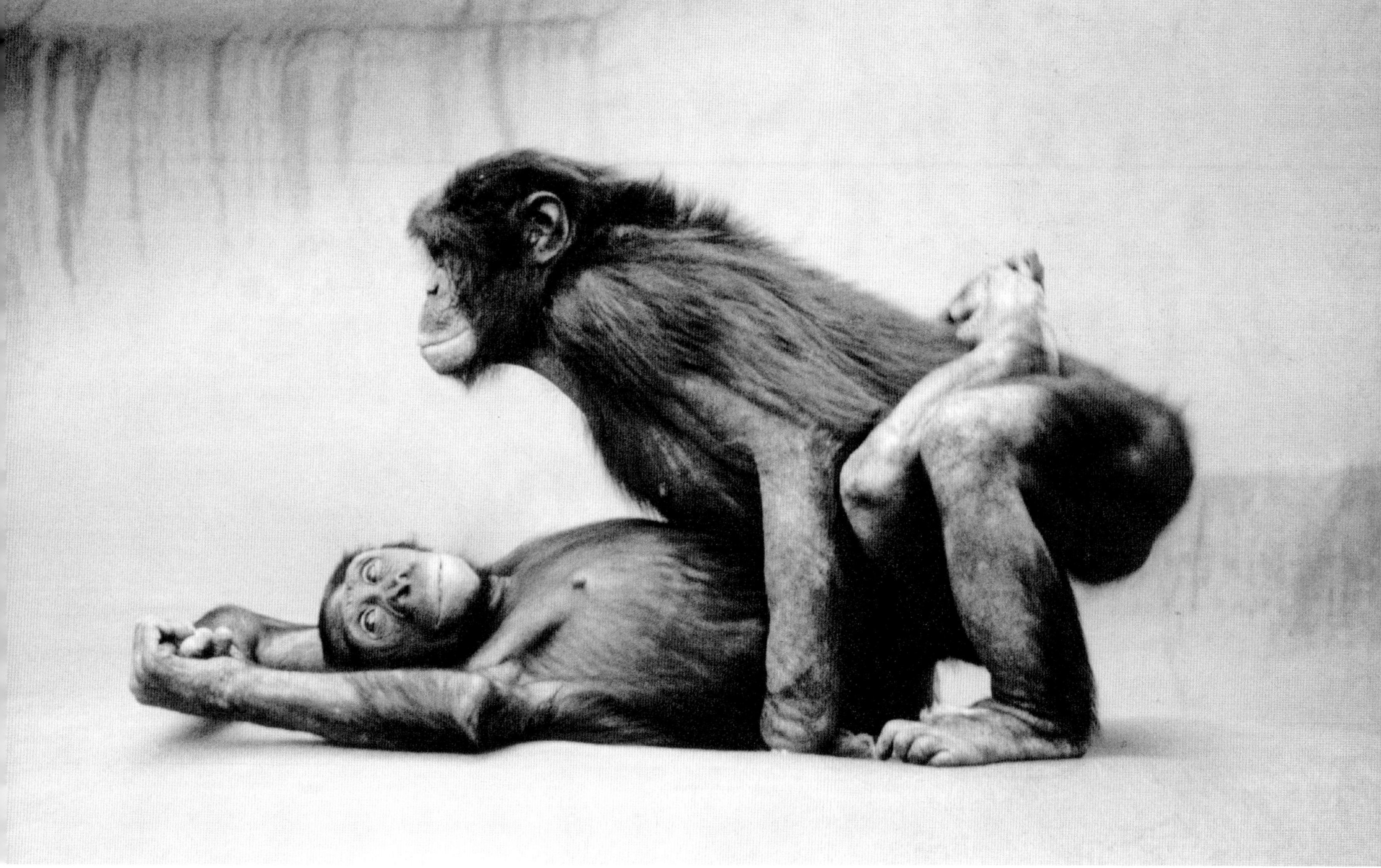

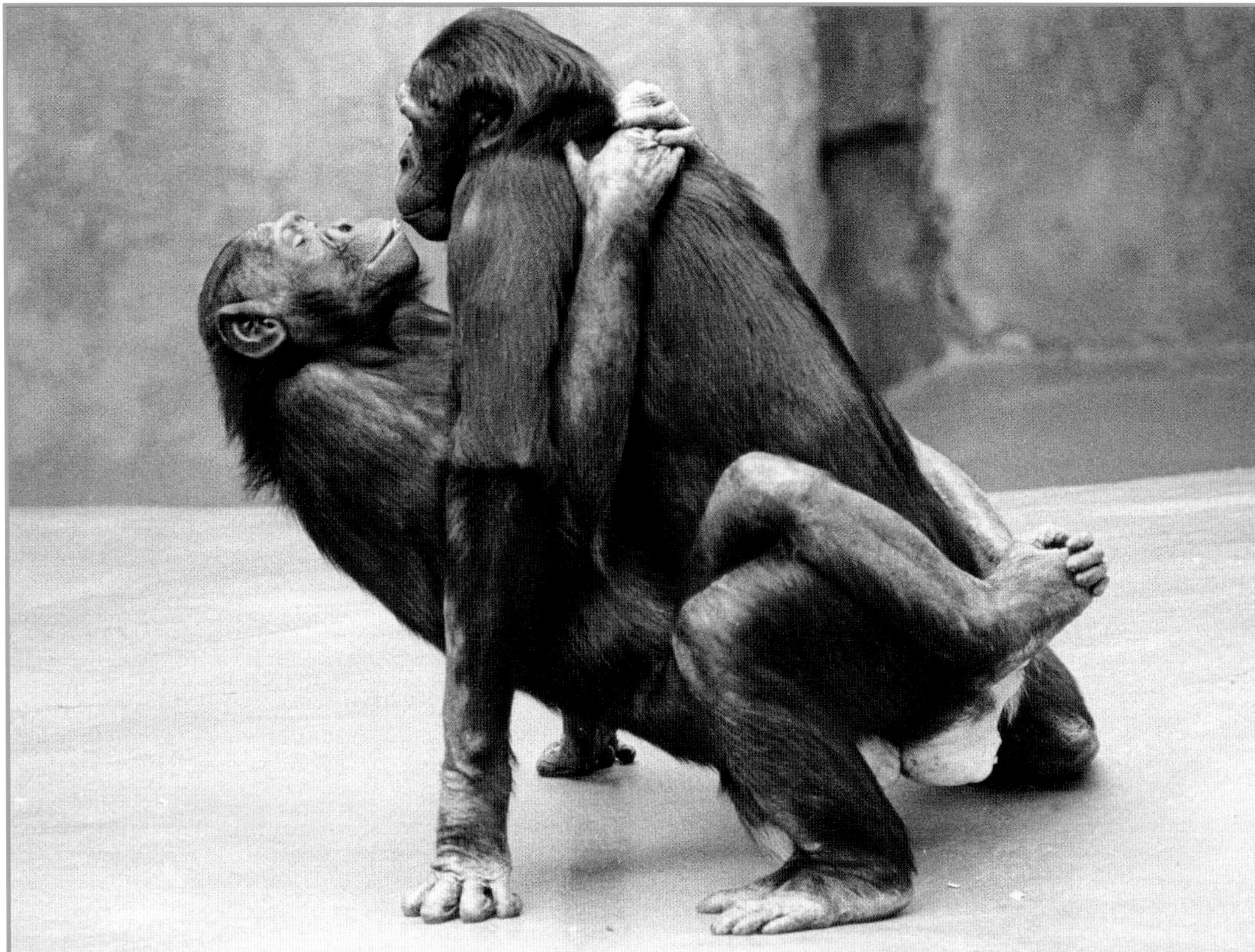

Sisterhood

Bonobos have sex in all possible combinations of individuals, not only between males and females but also in male-male, female-female, and adult-juvenile combinations. This widespread sexuality occurs partly because sex serves to reduce social tensions, and tensions can arise between any individuals. Another reason is that sex is a political instrument. The most important political ties in bonobo society are between adult females, who form powerful alliances.

Here we see sex between two fully grown females at the San Diego Zoo. The small photo shows an invitation between females, the large photo actual genito-genital (GG) rubbing. In this pattern, observed both in captivity and in the wild, females rub their swellings and clitorises together while adopting the position of a mother-offspring pair. One female clings to the other the same way that an infant clings to its mother. In this position, they engage in genital rubbing, which helps them cement a close female bond. These female bonds translate into alliances that allow bonobo females to collectively dominate the males.

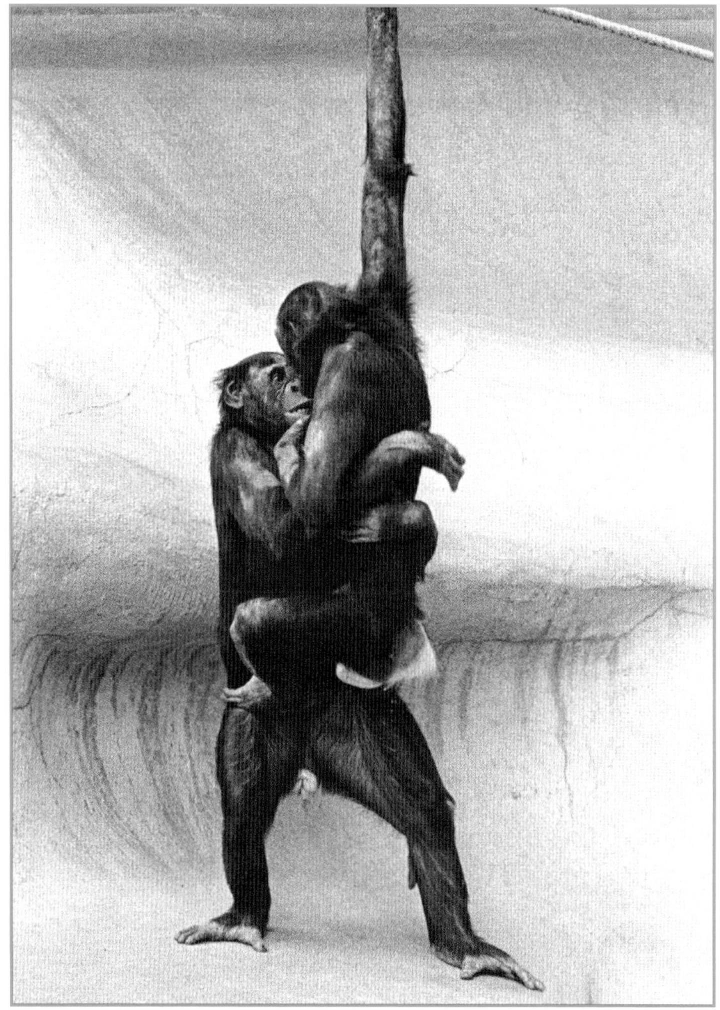

Big hair

While I was visiting the Western Ghats, in the steamy, tropical zone of southern India, my guides kept talking about LTMs—as in "LTMs here and LTMs there." They were referring to the lion-tailed macaque, one of the rarest, most endangered primates in the world. I would have had no hope of seeing these monkeys without the help of people who knew where to find them. Scientists of Mysore University have patiently habituated LTMs to the presence of humans, allowing a close-up view of the best head of hair in the primate world. The monkeys are not large, but they look impressive with their lion manes (with regard to their tails, the comparison with lions offered by their name is less apt, I feel). This adult female watched us curiously from a tree before descending to the ground, offering me just enough time to take a picture of her face. She then rushed over the ground to a distant tree, where a male was breaking open some giant jackfruit.

I was amazed by the way these monkeys travel in widely spread groups, often losing sight of each other, but staying in touch with calls. They always seemed to know where everybody was.

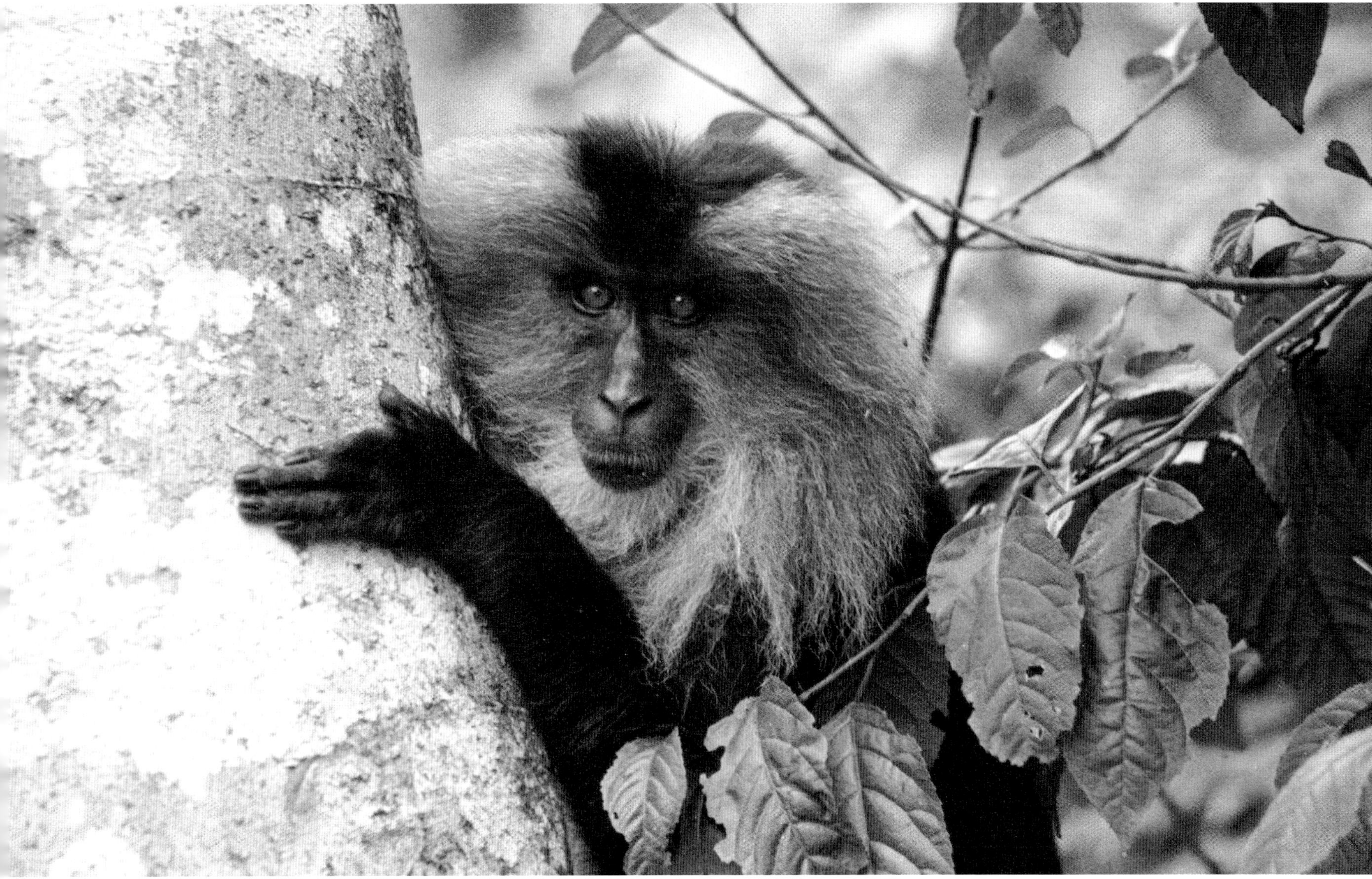

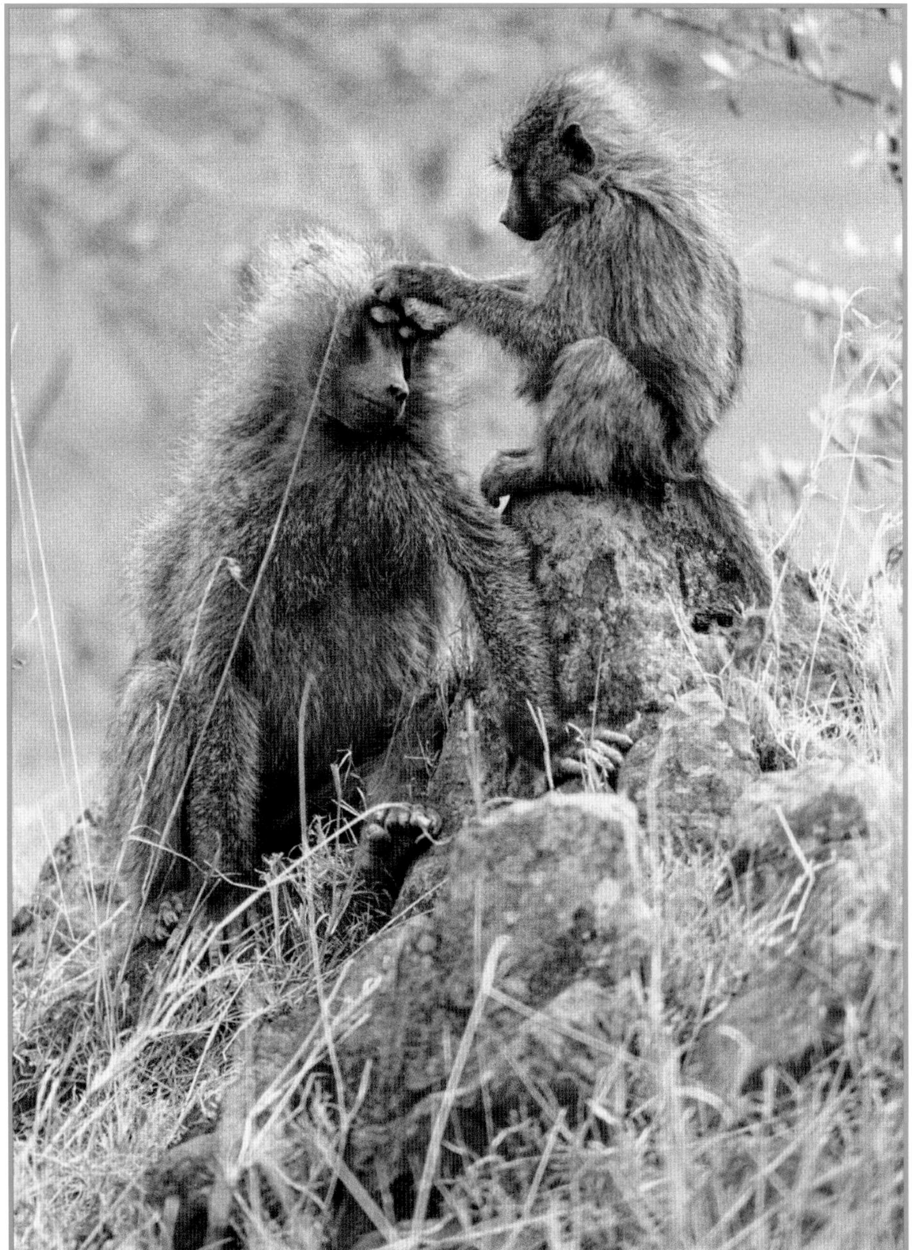

King of the mountain

It was fascinating to see how these two baboons arranged themselves, the female sitting next to the rock and the juvenile climbing onto it, before they began grooming. They sat down so smoothly in this "barbershop" arrangement that it was as if they had planned the positions ahead of time. Possibly they had groomed this way many times before.

I must add that the juvenile's being seated higher than the female does not reflect relative status in this society. People often think that sitting higher must indicate superior rank, perhaps because humans do arrange people in a hierarchical fashion, such as when we install a high table for dinner or seat kings and queens on thrones. This is a symbolic arrangement typical of humans, however: in monkey society sitting above or below someone else has no such significance. Not knowing these two baboons on the Kenyan plain, I do not know their relationship but assume from their relaxed attitude and intimacy that the juvenile was the daughter of the groomed female.

Duck face

The so-called duck face is the most endearing facial expression of the bonobo. The lips are pressed firmly together around the corners but left slightly apart in the middle, resulting in a trumpet-shaped lengthening of the mouth. The expression, associated with grooming, is shown by both groomers, as in the large photo, and groomee, as in the small photo. The face signals concentration on the grooming task as if to make clear to the partner that this job is taken very seriously and greatly enjoyed.

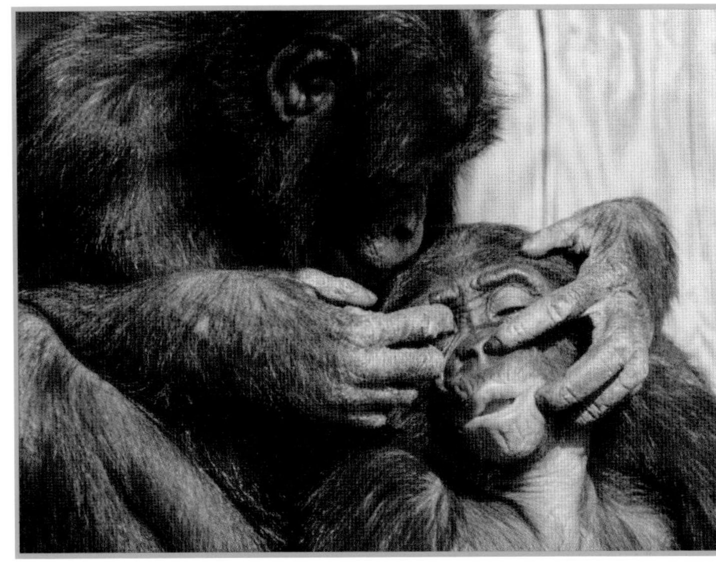

Because facial expressions are supposedly tied to emotions, they are thought to be universal species characteristics. Yet this expression may be unique to the bonobos at the San Diego Zoo. It has never been reported for any other bonobos nor for any other apes. The duck face may be part of the San Diego Zoo's grooming tradition, which also includes hand clapping (page 67). If so, it would be the first and perhaps only primate facial expression to show cultural variation.

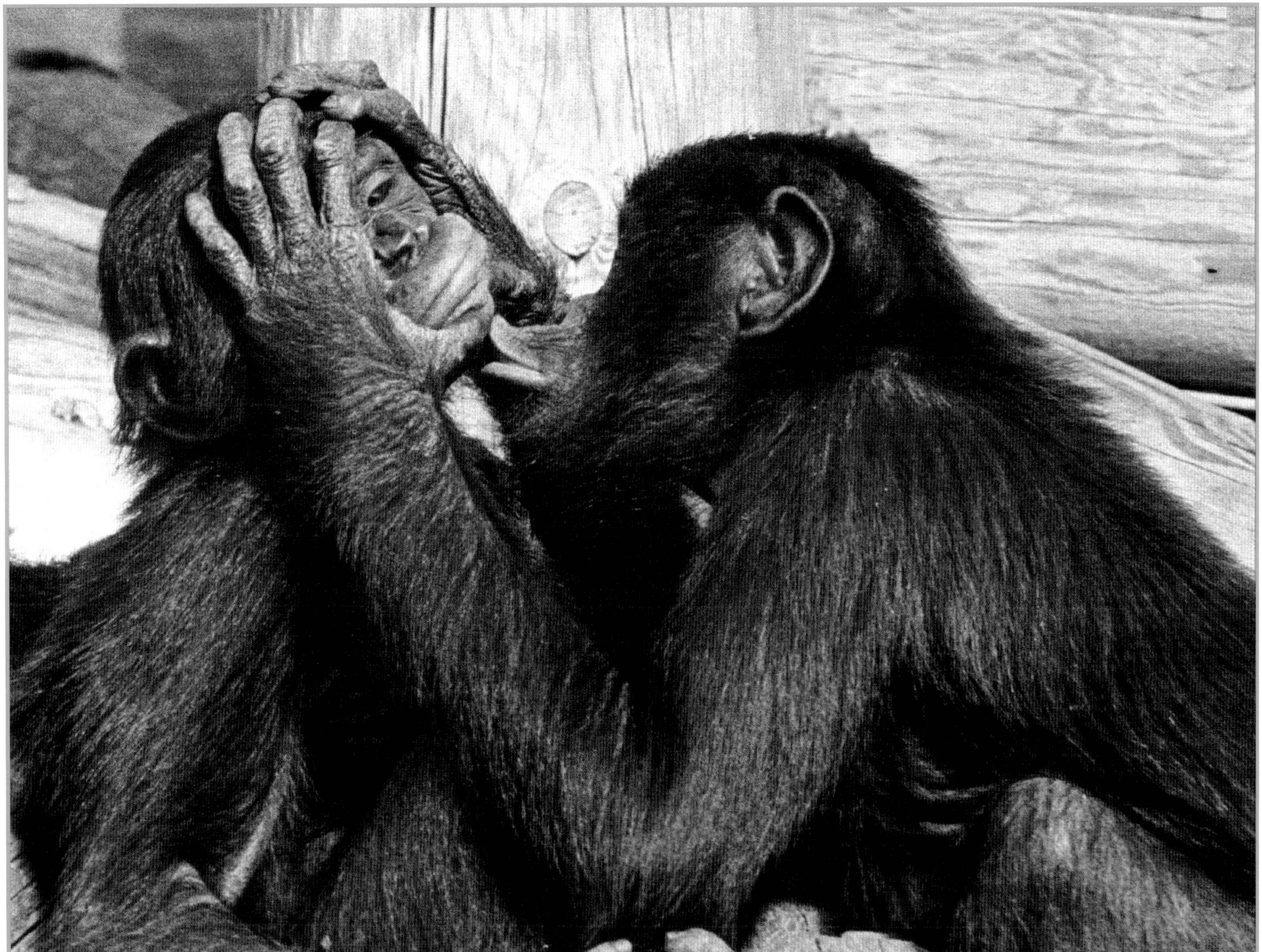

Audible grooming

Not to be confused with the chimpanzee's hand-clasp grooming (pages 68–69), hand-clap grooming is a unique tradition of the bonobos at the San Diego Zoo. It has not been observed in any other ape group, captive or wild. One bonobo claps her hands a couple of times, then starts grooming the other's face, after that alternating between clapping and grooming. Sometimes these bonobos clap their feet together, since they use their hands and feet interchangeably. This makes the San Diego Zoo the only place in the world where one can actually *hear* apes groom. When new individuals are added to the colony, they pick up the habit in about two years.

The clapping gives the impression of a ceremonial opening of the grooming contact. One individual walks up to another, claps her hands or beats her chest with a few rhythmic taps, then reaches out to the other, who by then has already adopted a grooming posture because it is obvious what will come next. Including also the duck face (pages 64–65), the bonobos have thus surrounded grooming with rituals that, I assume, enhance the pleasure and predictability of these contacts.

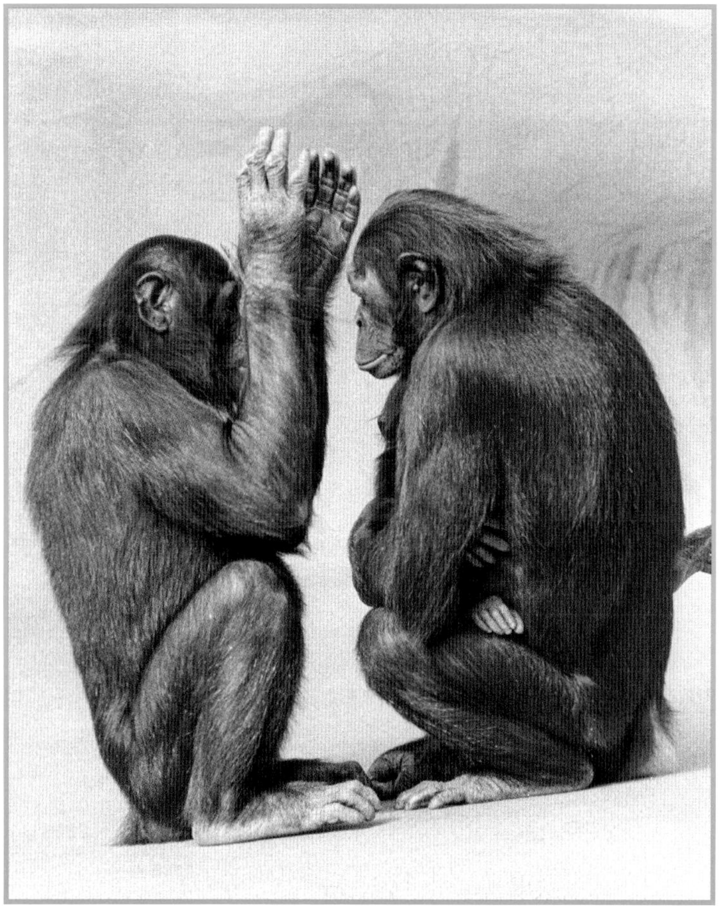
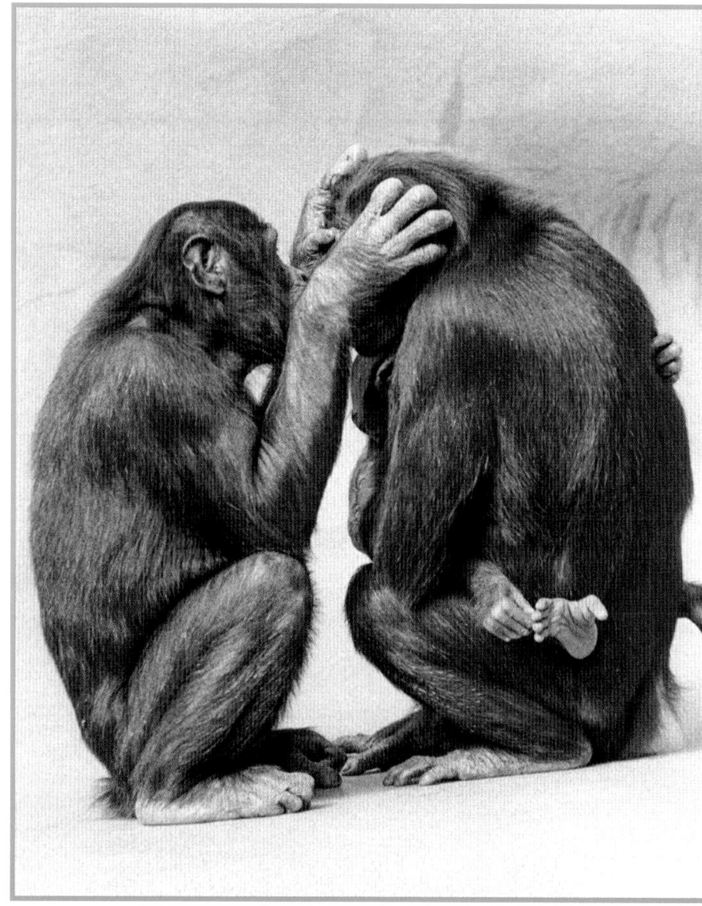

Grooming schools

Each community of chimpanzees is set apart from others by its special set of traditions, also known as its "culture." No other captive colony in the world shows the hand-clasp method of grooming, a characteristic of one of our groups at the Yerkes Primate Center Field Station. Two apes grasp each other's hands and lift both arms above their heads in an A-frame arrangement. With both hands in the air, they groom each other's armpits with their free hands. The hand-clasp gesture promotes intimate, mutual grooming. It probably serves to signal friendship. In our colony, the custom was introduced by a female named Georgia, and we were lucky to follow its spread from the start. Now, ten years later, it is practiced by all of the adults.

Hand-clasp grooming has also been noticed in a number of natural chimpanzee communities in Africa. In the wild, too, this behavior is cultural in that different communities may act differently. Of two communities that live virtually next door to each other, one may show the pattern every day, whereas the other may never show it.

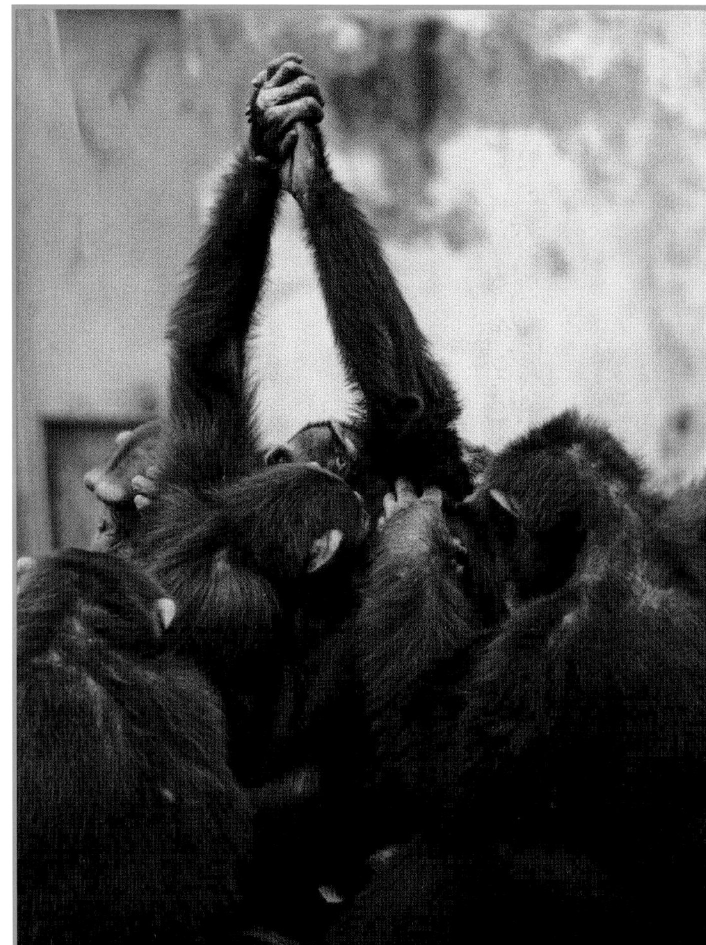

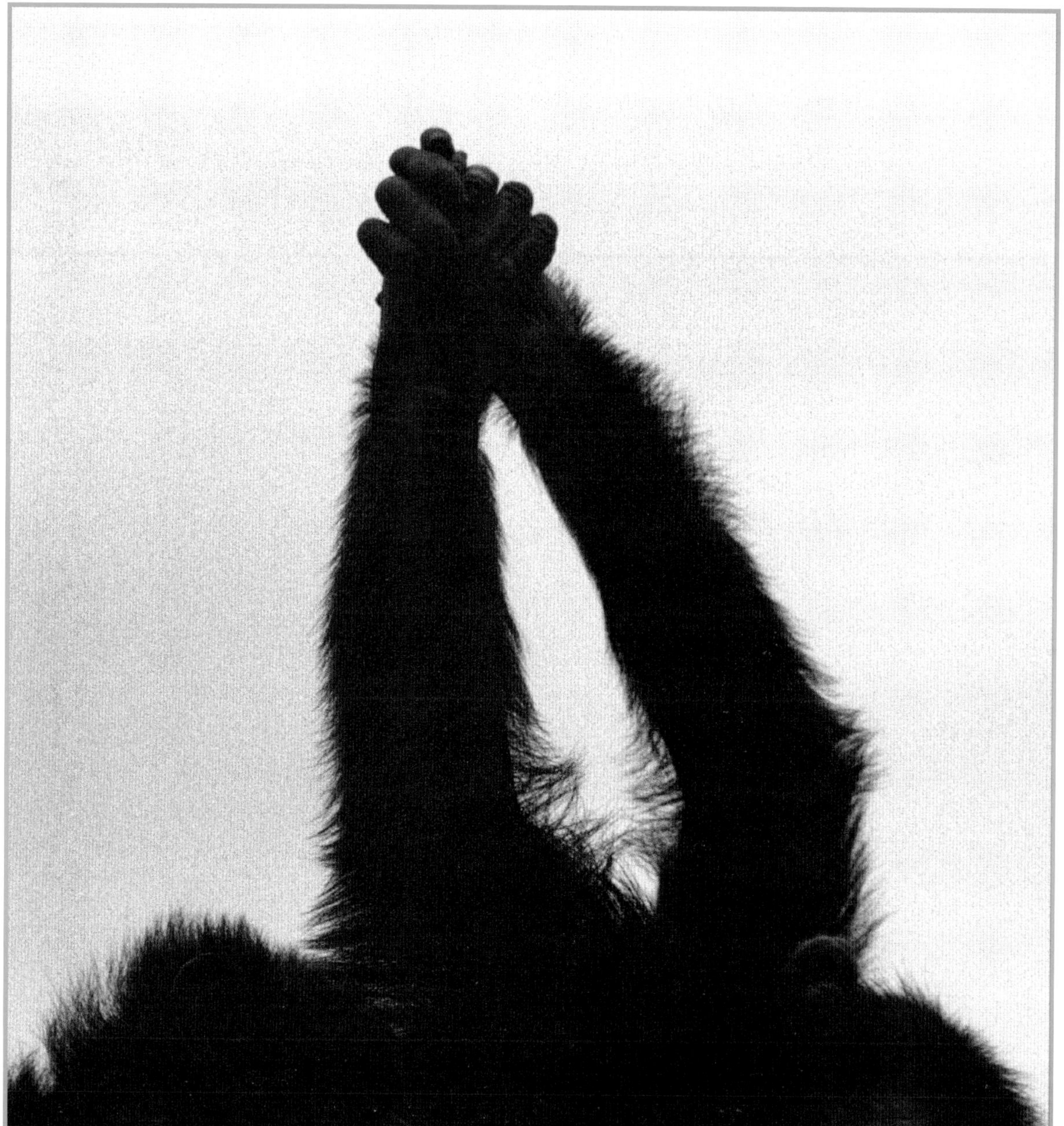

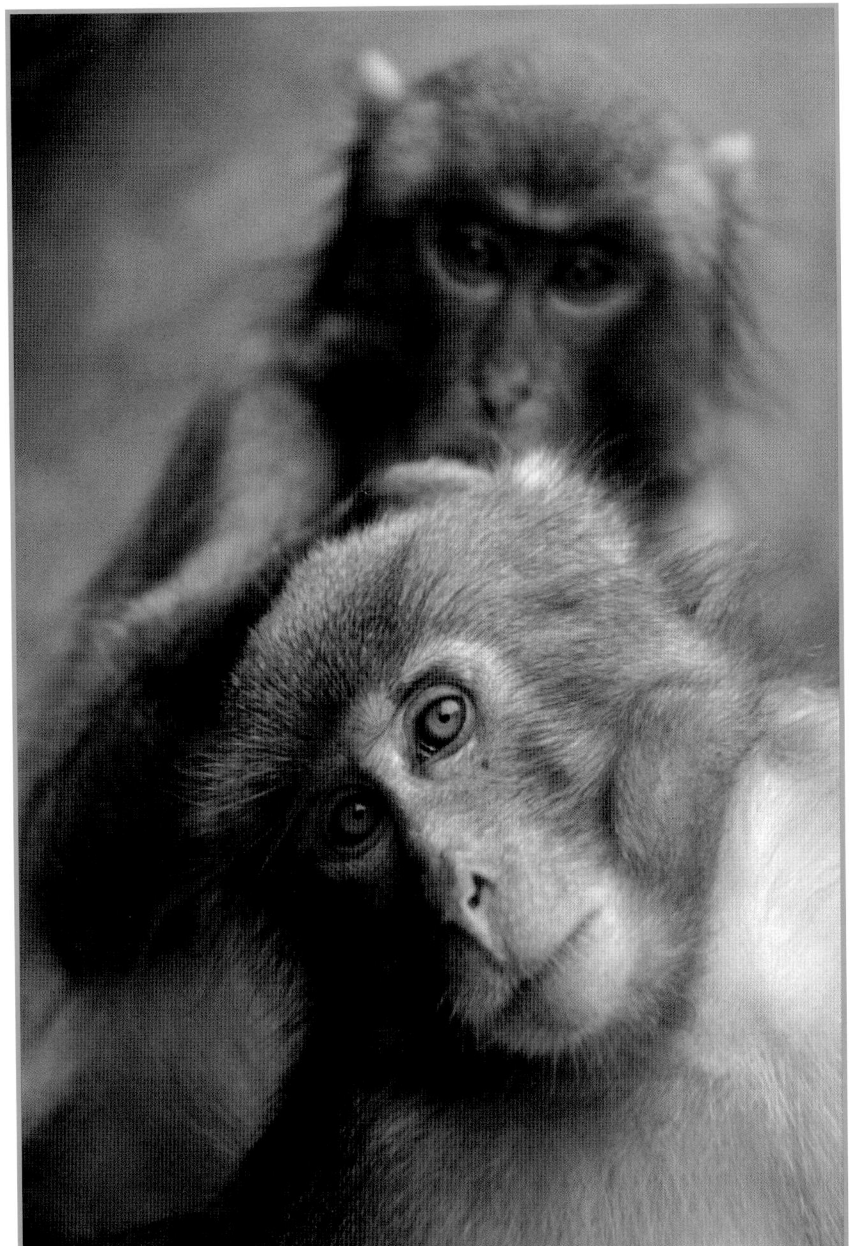

Koshima pioneers

Compared to mainland monkeys, the Japanese macaques on Koshima Island are undersized. They are only three-quarters normal size. The tiny island is estimated to produce natural foods for about thirty monkeys. But through provisioning of sweet potatoes and wheat, the population has grown to about one hundred. When scientists sharply curbed provisioning on the island twenty-five years ago, the monkeys quickly depleted resources to such a degree that the bird population, with which they compete, declined, and the monkeys themselves began to drop in weight and size. The Koshima monkeys are famous as the textbook example of animal culture: they developed potato-washing behavior in the 1950s and still show this behavior on the handful of occasions every year that scientists empty bags of sweet potatoes on the beach (pages 146–47).

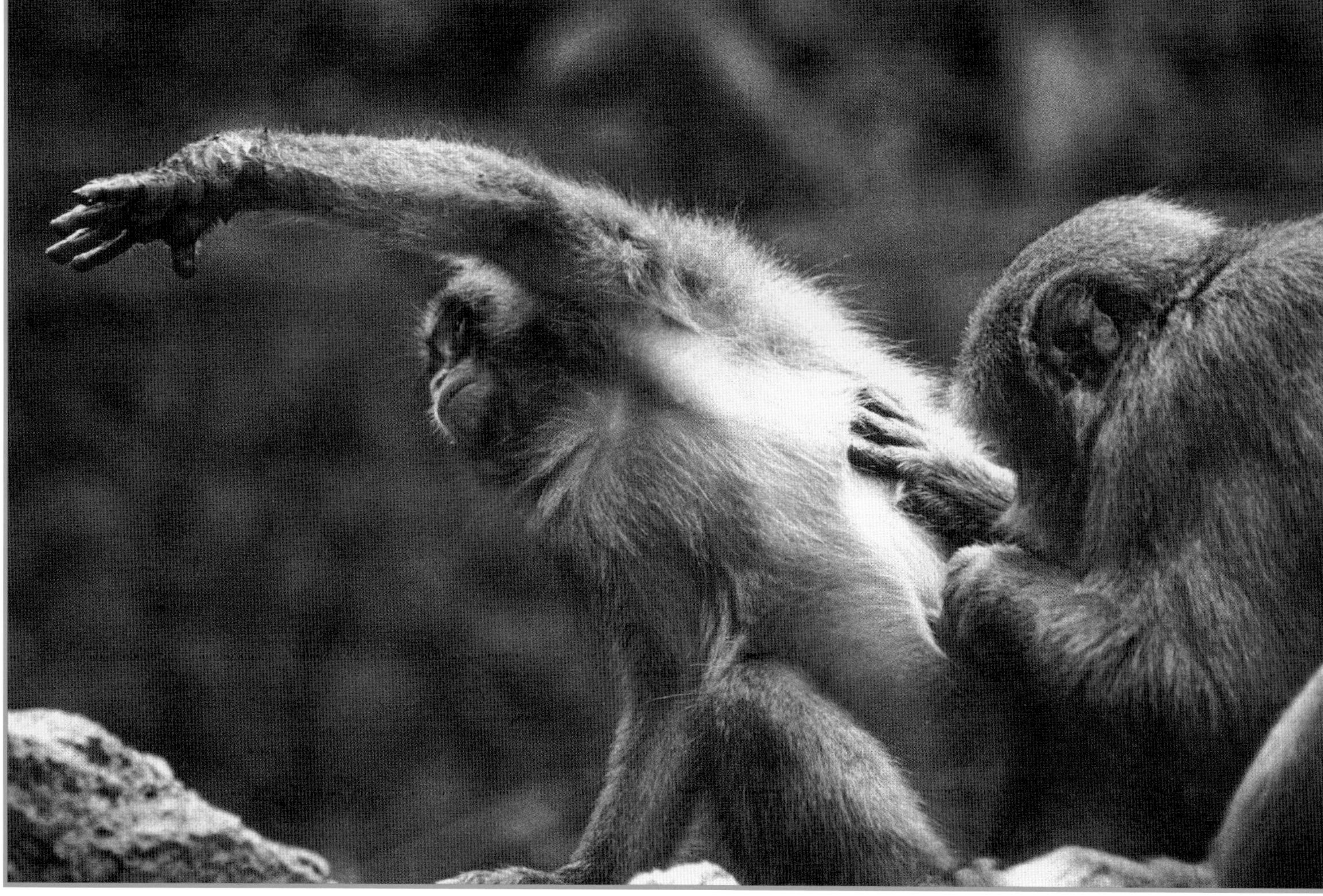

The pleasure of touch

For monkeys and apes, being groomed by a mother, sister, or close friend is a soothing experience. This is why they show so much abandon and trust when presenting themselves for the treat. They turn and stretch while their partner is busy cleaning and combing through their hair, removing and eating salt crystals, ticks, and lice eggs. But even if hygiene is the original function, most grooming has more to do with attention and friendship than with removing ectoparasites. Studies have shown that a monkey's heart rate drops while it is being groomed, thus indicating relaxation. Before the age of showers and shampoo, grooming was probably far more common in human society than we realize. We still have massage parlors, spas, and beauty salons that cater to this need. In primates, grooming serves as the glue of society. Its beneficial, enjoyable nature turns it into an ideal social "currency" for all sorts of transactions. Grooming is used to foster bonds as well as to obtain favors: the Japanese monkey posed with its arm in the air in Takasakiyama, Japan, may remember the groomer's service and repay with a return favor when the occasion arises.

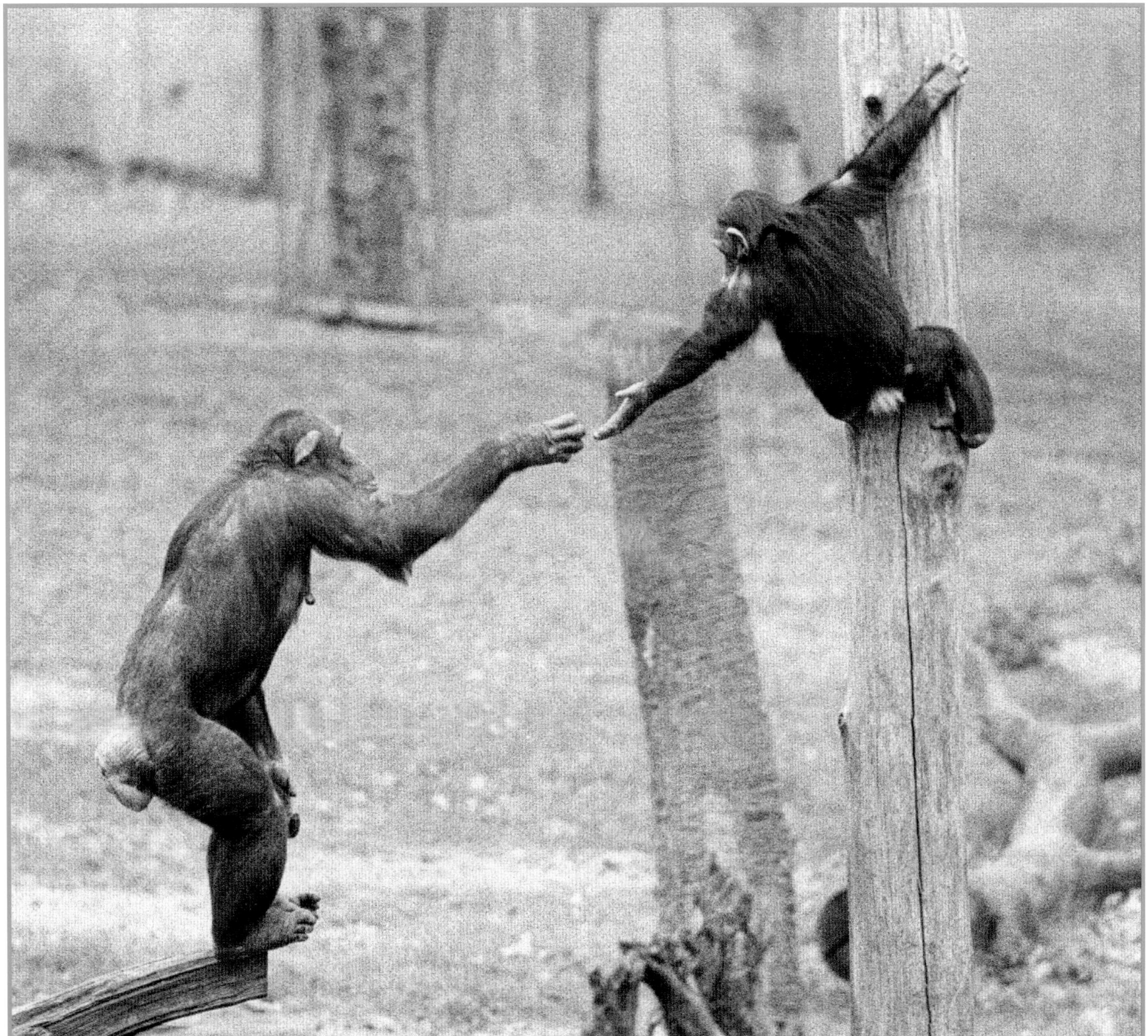

Empathy

Empathy permits one individual to relate to the emotions of another. Many animals show the simplest form, in which individual A gets upset when B is upset, much like a human infant who cries when it hears another infant cry. More advanced empathy develops quickly in children, however, and is also found in chimpanzees. For example, the juvenile male in the small photo has rushed to a screaming adult male, who was defeated in a fight, to wrap an arm around him and reassure him. Consolation is common in apes yet absent in most other animals. Chimpanzees also show targeted helping, which requires them to evaluate not only the emotions but also the situation of another. Thus, the mother chimpanzee in the large photo hurried to her young son after he had uttered only one brief scream of distress. She is stretching out a hand to help him out of a tree: he panicked trying to descend on his own.

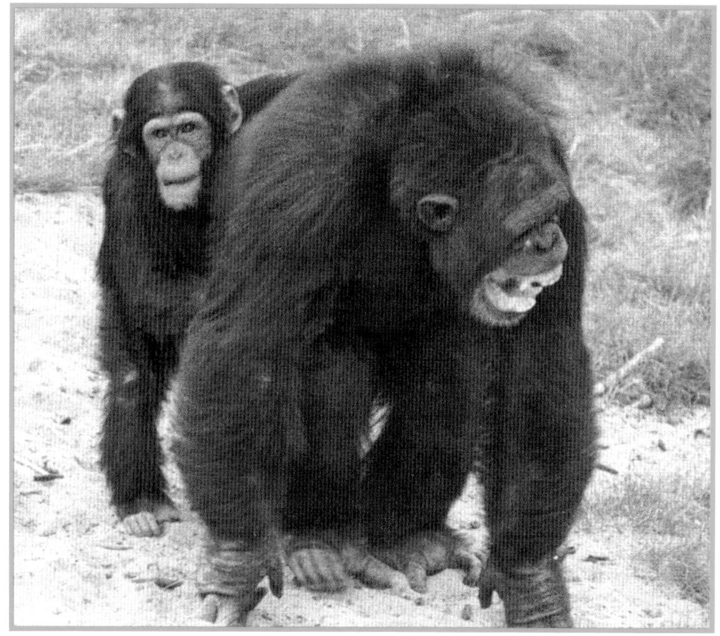

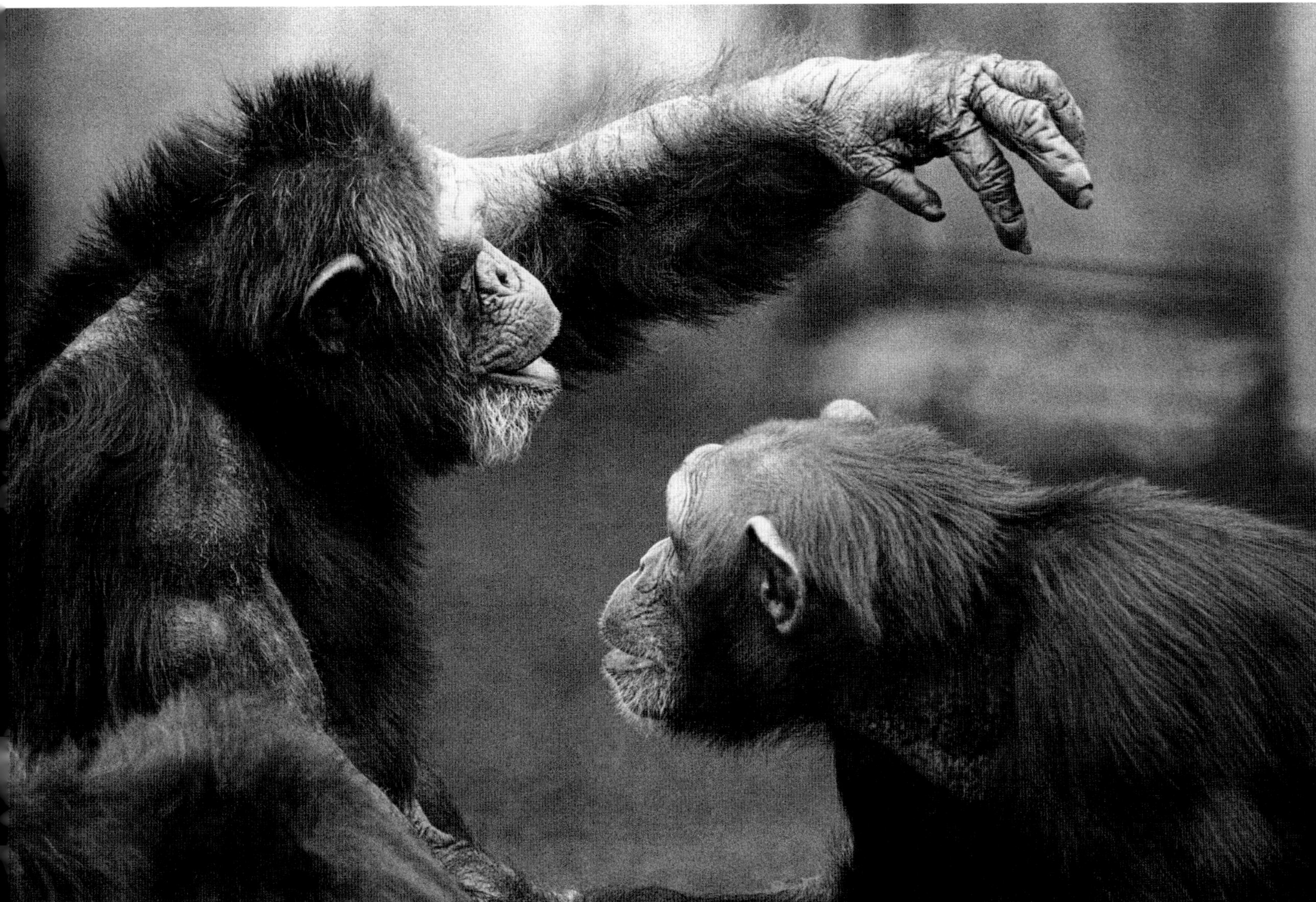

Delicate dance

Dancing is a dialogue in which the body movements of each partner affect those of the other. That same kind of back-and-forth communication happens with chimpanzees. Here Georgia approaches the dominant male, Jimoh, in a ritual that I saw repeated several times per day one summer. She would approach rather submissively, looking up at him from a lowered position. Occasionally, Jimoh—who must have gotten tired of her insistent attentions—failed to respond. This would call off the dance or rather the dance invitation. But usually Jimoh saw her coming from afar and nodded his head in approval or, as here, welcomed her by lifting up an arm in a patriarchal gesture. Georgia would then seek eye contact, sometimes pouting her lips at Jimoh. This facial expression, which reflects a desire for affection (and is typical of infants wanting to nurse), is on Georgia's face in the small photo on the left. The same photo shows that Jimoh is ahead of her, because he already closes his eyes and sticks up his chin, knowing that she will place a big, wet kiss on him, which Georgia does in the small photo on the right. The kiss completes the greeting sequence.

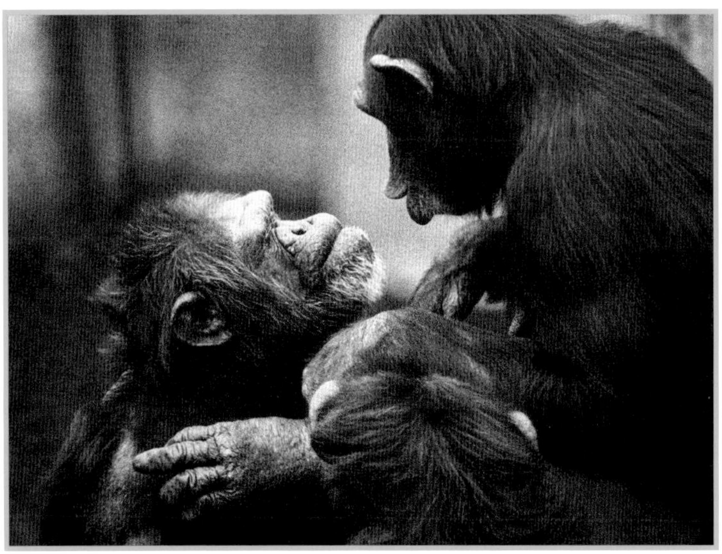

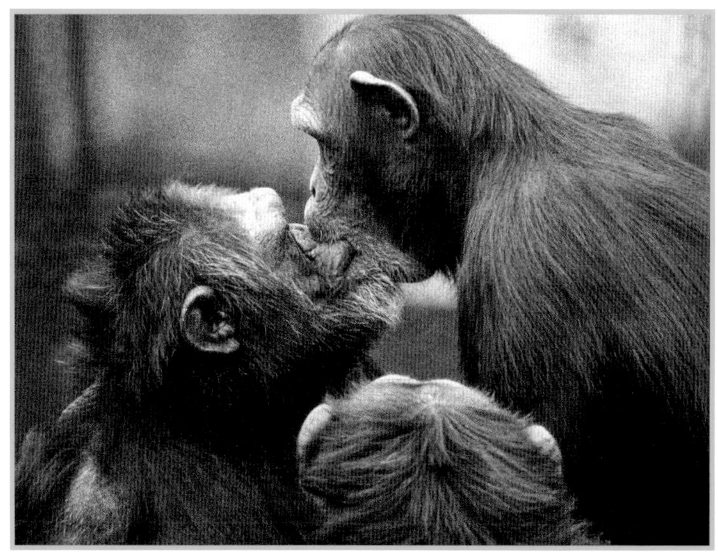

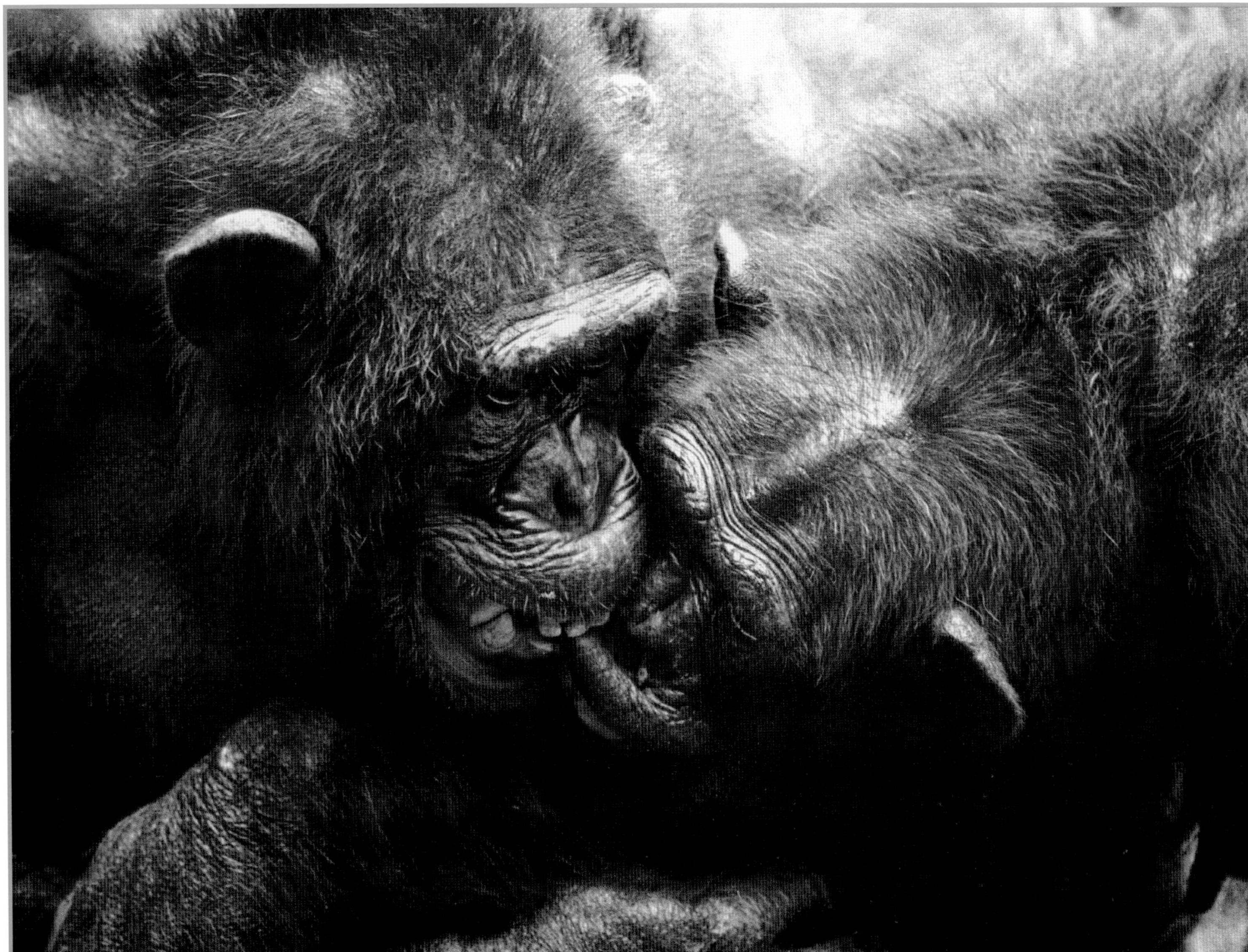

Good kissers

This kiss was triggered by the end of a major group conflict in which these two individuals were involved, even though they had not been directly opposed to each other. "Celebrations" of joyful events by means of kissing and embracing pave the way for peaceful coexistence. Chimpanzees have many ways of kissing, from softly gnawing on somebody's arm or shoulder all the way to excited mutual lip biting with bared teeth as pictured here (large photo) between an adult male (left) and female. Most of the time, kissing is a kind of reassurance, such as in a greeting after a separation or a reconciliation after a fight (page 77).

The chimpanzee's kiss does not have the sexual meaning that a human kiss often conveys: the chimpanzee's kiss is a "platonic" contact. This does not apply, however, to bonobos, who French-kiss and hence have turned the same contact form into an erotic affair, with mouths wide open and tongue interplay (small photo).

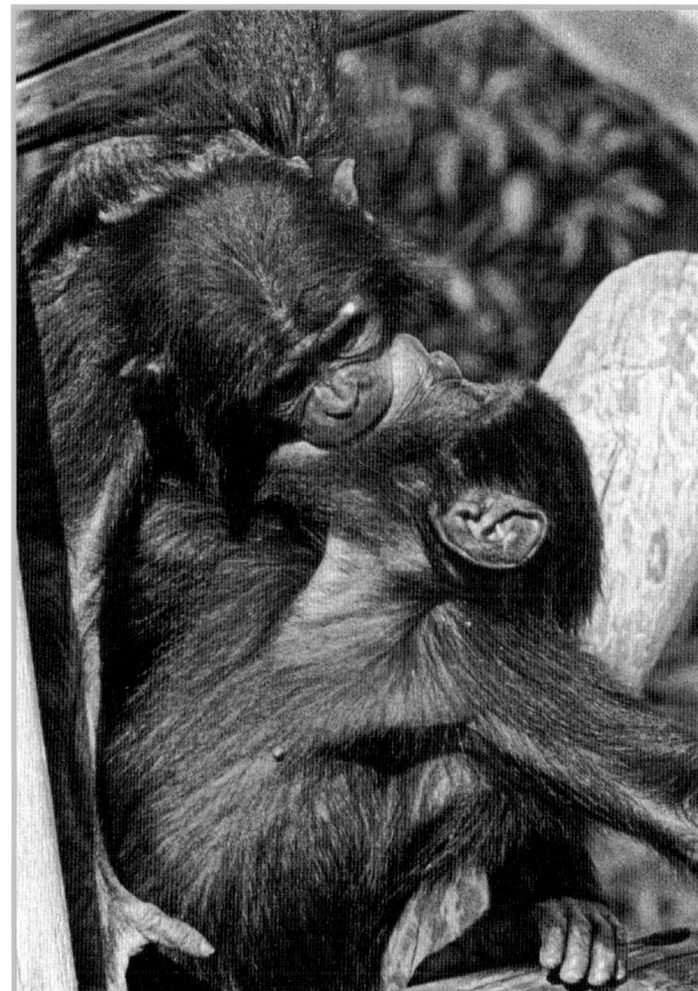

Dog-faced

Baboons have an exceptionally prominent snout, so much so that they used to be known as dog-headed or dog-faced monkeys. Young adult males, such as those in the large photo, are built like fighting machines (see also page 82): they are twice the size of females (see small photo) and have canine teeth as formidable as a panther's. Roaming dry savannas and woodlands, baboons used to be popular with anthropologists, who regarded them as the best living model of our ancestors. Part of the appeal for comparison was that these terrestrial primates are adapted to the sort of ecology that proto-hominids must have faced after they descended from the trees. The baboon model was largely abandoned, however, when it became clear that a number of fundamental human characteristics are absent or only minimally developed in these monkeys, such as cooperative hunting, food sharing, tool use, and territorial warfare. Chimpanzees offered better parallels. Yet in the meantime, we have learned enough about these rambunctious monkeys of the African plains to recognize that a baboon troop is one of the most intricate animal societies on earth.

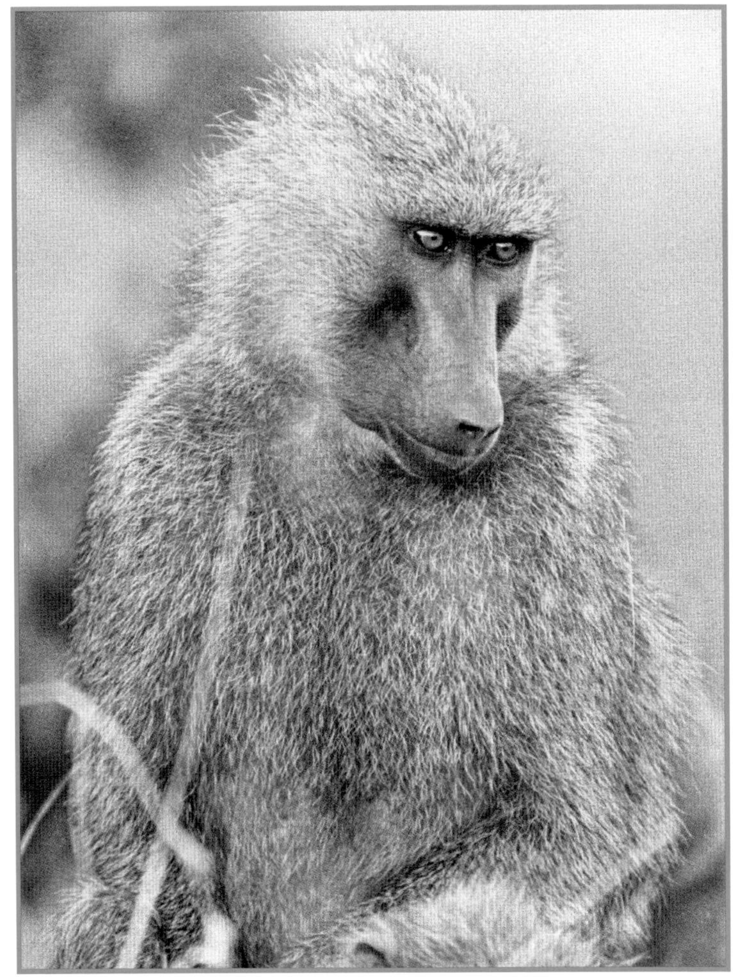

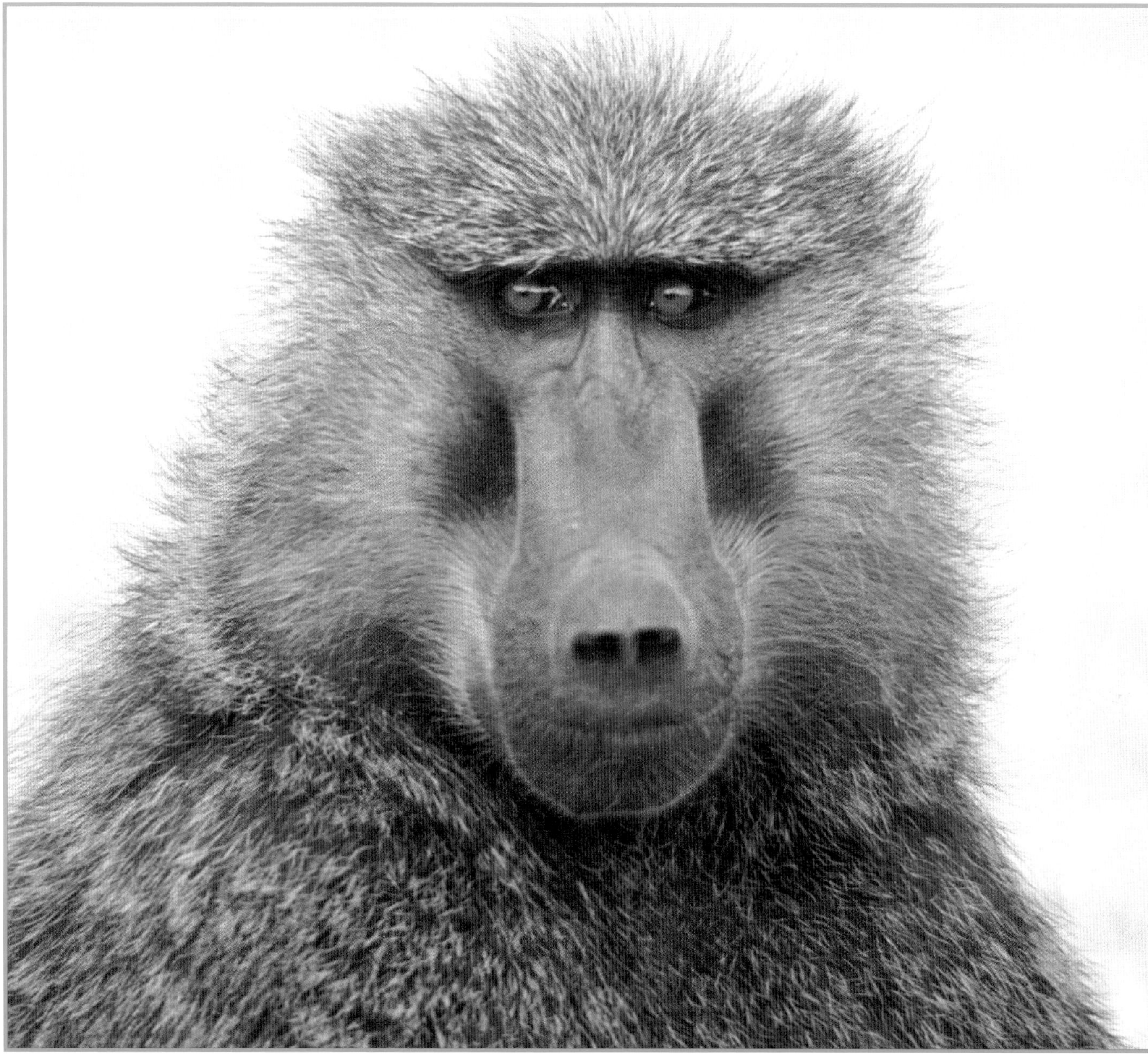

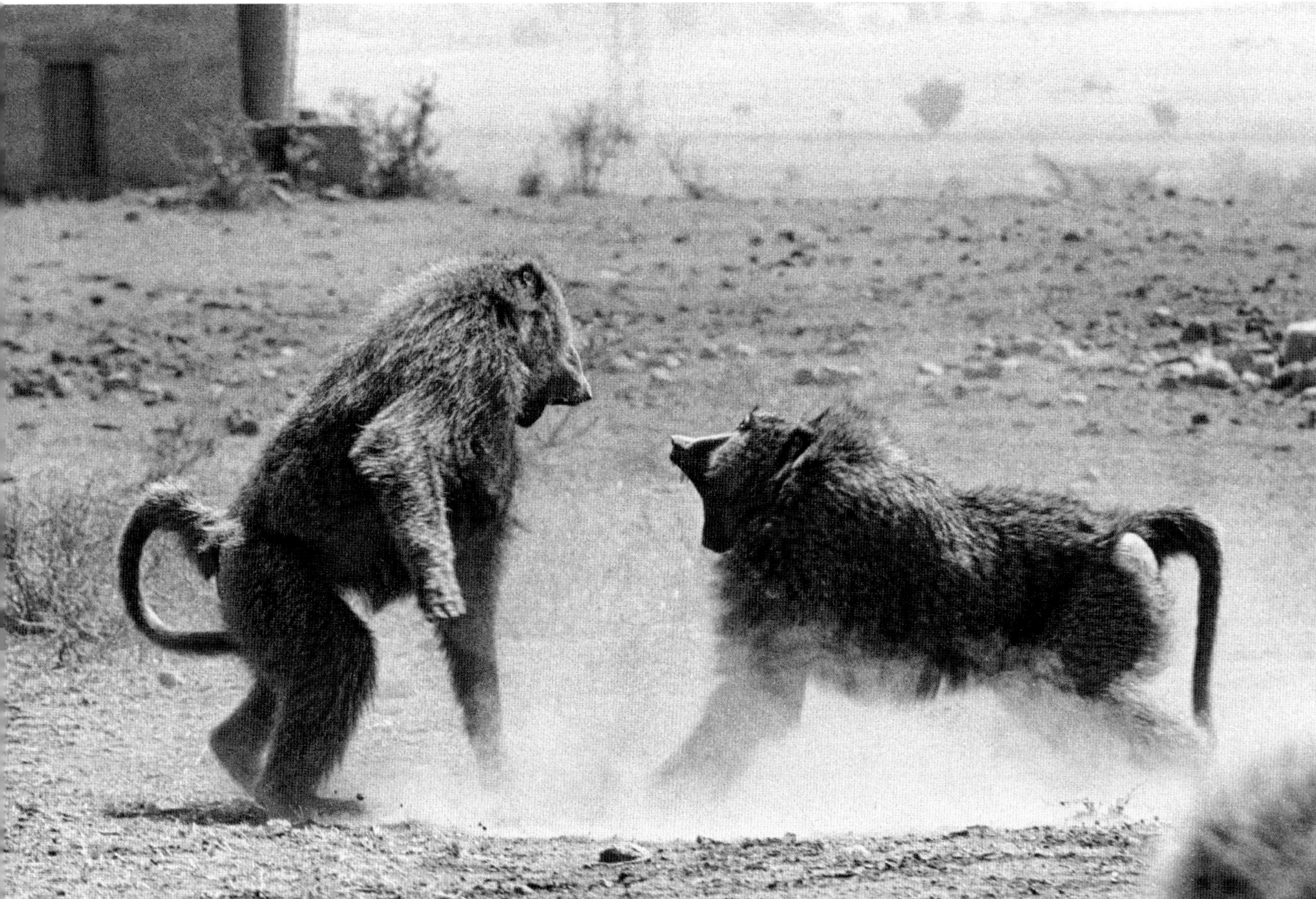

Face-off

Two wild male baboons go at each other in a whirlwind of dust. One male stands almost upright and the other's shoulder hair stands up, while the two slash at each other with their outsized canine teeth. The scene, which took place on a ranch in Kenya, was over within seconds. Despite their short duration, these fights can produce terrible injuries: one canine slash can cut the entire length of a face. Fighting males cannot afford to turn their backs. They spar face to face in order to avoid getting bitten elsewhere, which explains the predominance of facial injuries.

Because these physical fights are spectacular, they give the impression of being all-decisive. In reality, they are only the tip of the iceberg of male rivalry. Males often test social relationships by other means. A male will intimidate another by walking very closely past him, to see if he flinches. Or one male will continuously follow another around, making him nervous by watching every move he makes from a distance of a few meters. This subtle war of nerves explains why fights often erupt over matters far too small to be the real cause.

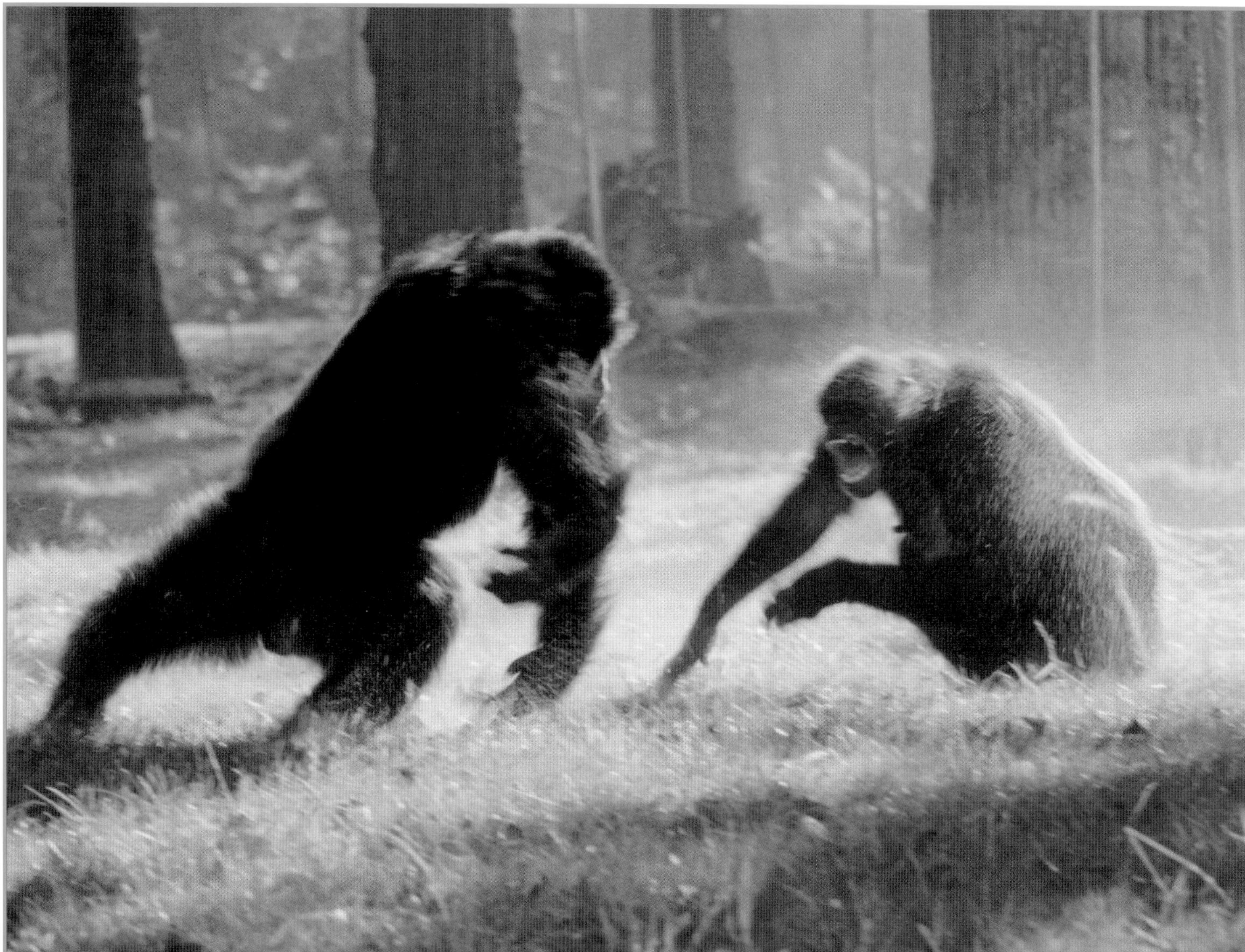

Limited dominance

Powerful blows by the male on the left have sent the female on the right tumbling through the sand. She screams in protest. Even though normally any healthy male chimpanzee dominates any female, this does not mean that he can attack her with impunity. Males usually fight females only with their hands and feet: use of their dangerously sharp canine teeth is limited to rare fights with male rivals. If a male on occasion does bite a female or attack her more intensely than usual, other females in the group disapprove. They bark with a "woaow" sound in protest, and their barks soon swell to a chorus. If this doesn't stop the assault, several females may band together to liberate and protect their unfortunate "sister." No male dares to resist such a rescue operation, because, unlike the males, females have absolutely no qualms about using their teeth. They chase the male until everybody is tired, teaching him that dominance over females is accepted only within certain limits.

Peace in the treetops

I vividly remember how this uniquely silhouetted photograph of a reconciliation was made, even though now more than twenty-five years have passed. Two adult male chimpanzees had chased each other around the large Arnhem Zoo enclosure, screaming at each other but without displaying physical aggression. Nikkie (on the right) had chased his rival, Luit, high into a dead oak tree, where the two finally gave up and calmed down. They had been sitting there panting for about ten minutes, and I was reviewing the whole episode with the keeper, who stood at the moat leaning on a rake. We often talked about our chimps in the way that others discuss soap operas. In the midst of our chat, Nikkie all of a sudden stuck out his hand as a peace offering to Luit. I scrambled to aim my camera and focus. I had barely taken this single shot when both rivals left their positions. They descended to the fork of the tree, where they kissed and embraced. Then they climbed together to the ground and groomed each other for the longest time.

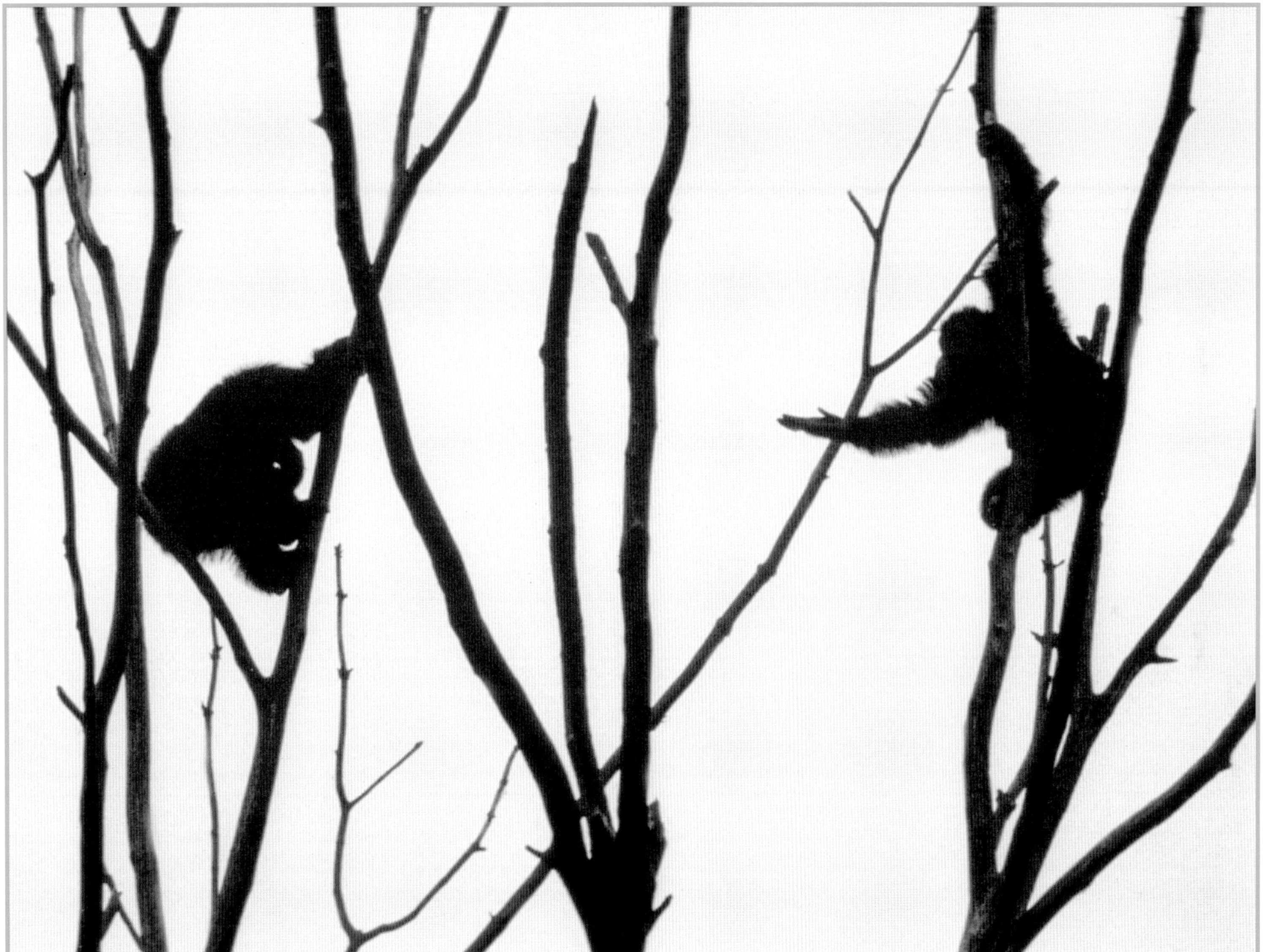

Family reunion

A nasty fight erupted within the "O-family" of our rhesus troop that involved two adult daughters of Orange, the alpha female. Sisterly feuds are the worst, especially high up in the hierarchy, where so much is at stake. Since one of Orange's daughters is likely to take over when their mother becomes sick or dies, the most dominant one has an interest in asserting her position. In this fight, Orkid (adult on the left) bit the older Omega (adult on the right). The aftermath was tense, but I was waiting for a reconciliation. Even though rhesus monkeys are not nearly as good at making up as some other primates, within the family reconciliation is common. In the competitive environment of a macaque troop, families need to stick together, and how could they do so without repairing damaged relationships? Orkid and Omega soon joined their mom, Orange (center), and all three of them huddled closely together while "girning" loudly (a friendly vocalization), even lip-smacking at each other's infants. As is typical of macaque reconciliations, eye contact was avoided. Soon the three were grooming, and peace was restored.

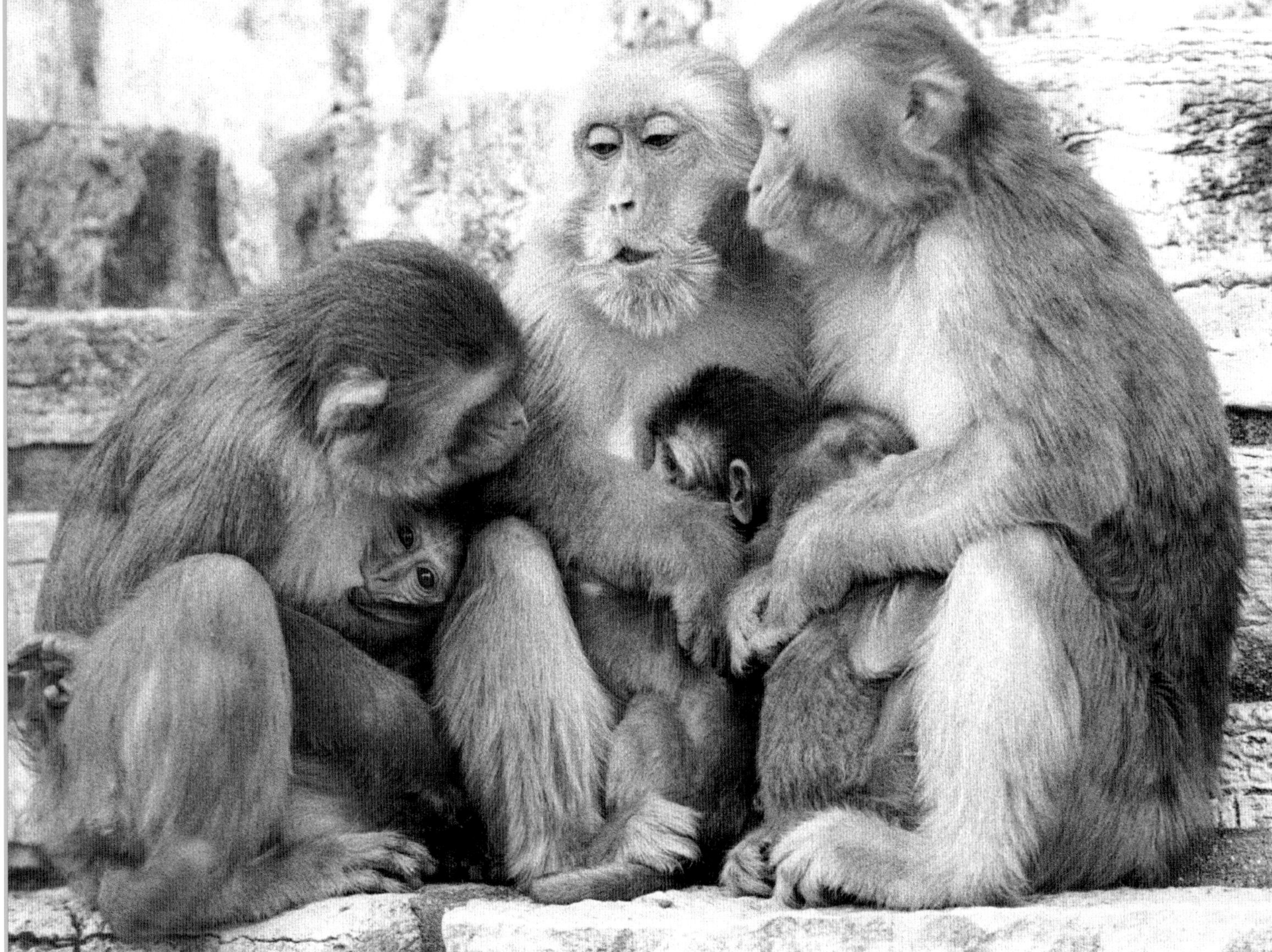

Symbolic punishment

A dominant female stump-tailed macaque has grabbed the arm of a juvenile, who, without fear, pushes his wrist straight into her mouth, letting her gnaw on it with a fierce expression on her face. To take punishment passively, reflecting agreement about who is boss, is not unusual among social animals. Stump-tails go through this kind of ritual all the time. Their "mock" bites never cause any injury; they probably do not even hurt. It looks as if the female is saying, "I could bite you. Perhaps I should bite you. But I won't." These status rituals signal mutual trust and agreement, underlining the social hierarchy without resorting to fighting. Stump-tail society is extremely close and tolerant, and the species can express status differences in ways that preserve rather than interrupt social cohesion.

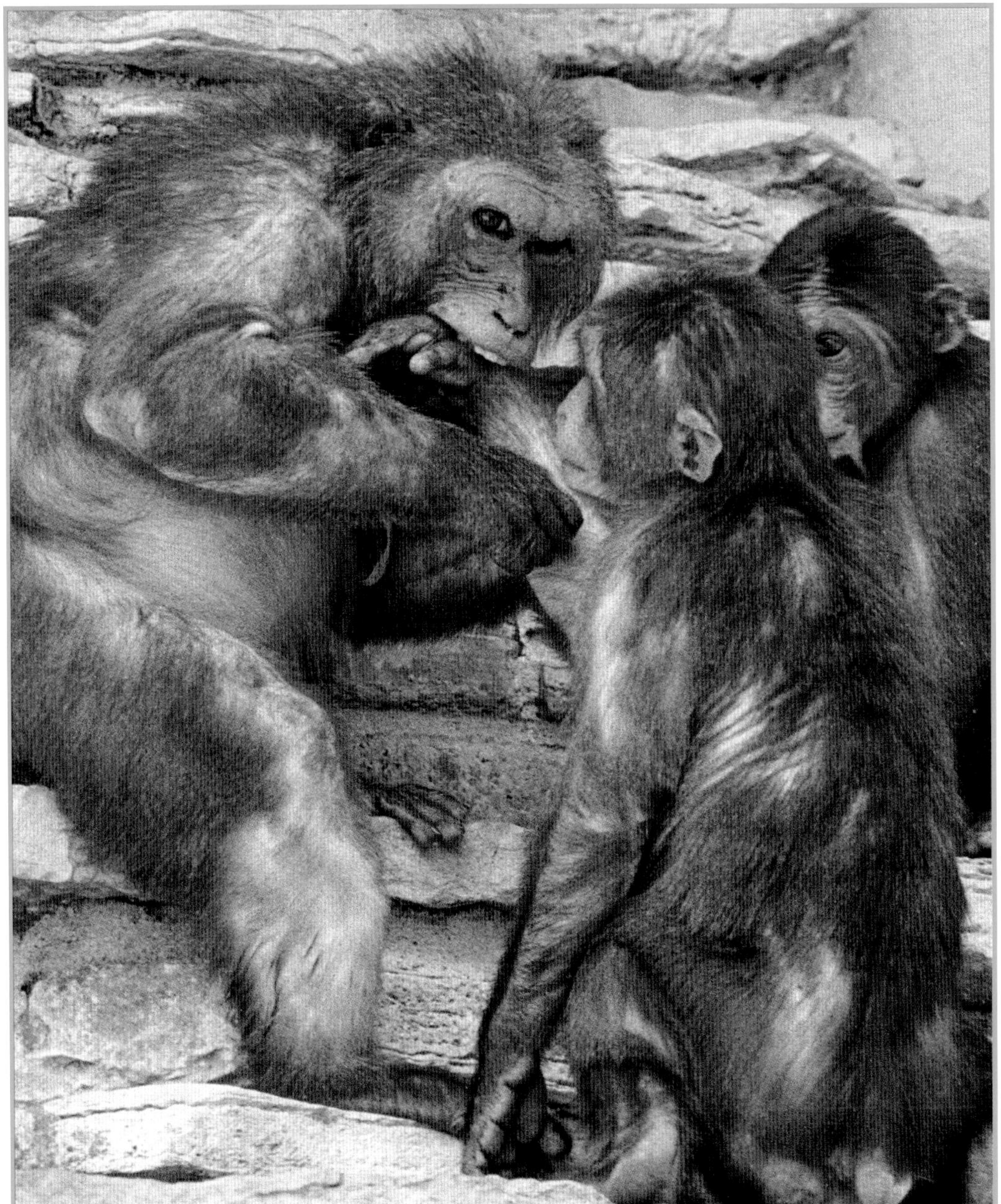

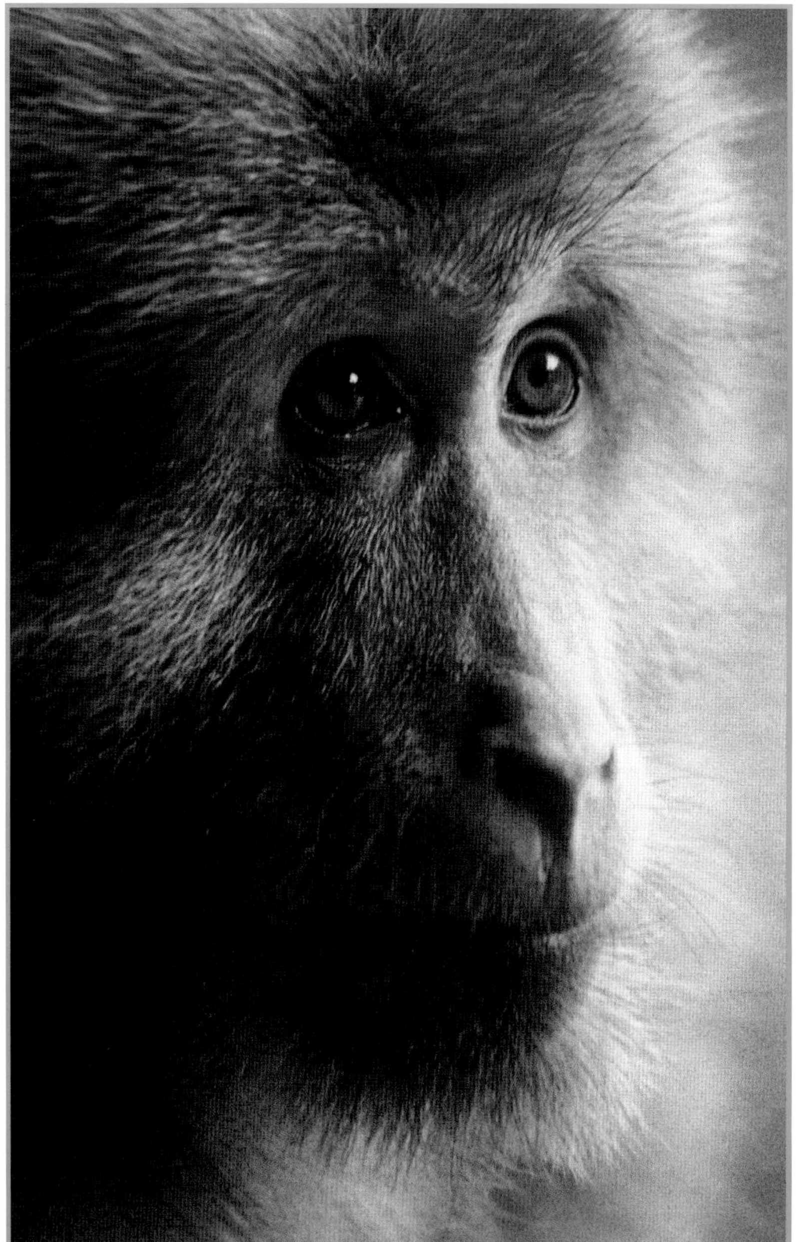

The awkward age

To be an adolescent male macaque is no fun. At this age his ties with maternal kin weaken, adult males in the group become unfriendly and intolerant toward him, and he will have to leave the natal group behind in search of a new home. This is not only a challenging step but a dangerous one: outside the group there are enemies and predators, and a new group rarely admits a newcomer without a fight.

About five years old, the male Tibetan macaque in the larger photo was so scared of the adult males in his own group (such as the one in the small photo) that he preferred to hang out with a bunch of visiting primatologists, knowing that our presence would buffer him from hostilities. In true macaque form, he coordinated his movements with us without ever making eye contact. We, in turn, scrupulously avoided staring at him: staring is a threat in macaque society. As a result, we became close, and the young male let me push my camera (from the side) into his face for the most intimate portrait of a wild primate that I have ever taken.

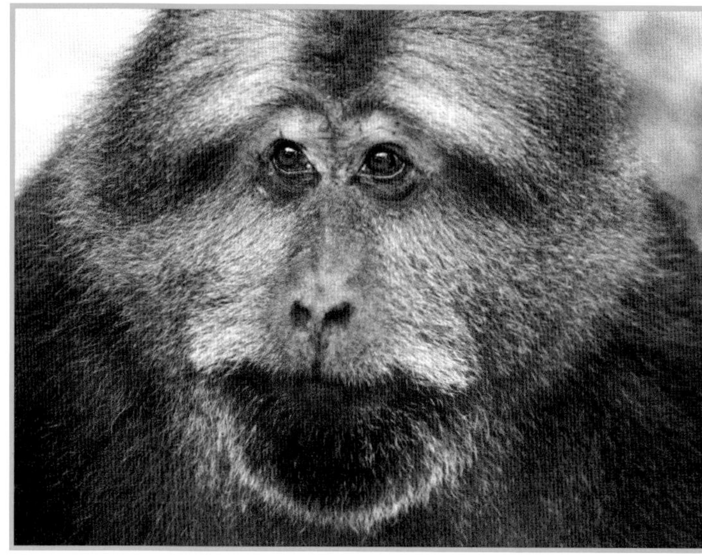

Pickpocketing

Being dominant in rhesus society has lots of perks. For example, a monkey of high rank can without impunity remove food from a subordinate's mouth. Doing so is made easier by the cheek pouches typical of macaques, in which they store unchewed food. A dominant female holds the head of a subordinate still while she checks out the contents of his pouches: if she finds something good, she will eat it straight out of his mouth! Whereas in all animal species social dominance implies inequality, it rarely goes as far as seen here. Rhesus monkeys simply are not your average primate: they have been called the chickens of the primate world because of their obsession with the pecking order. If we consider a range of dominance "styles," from egalitarian to despotic, rhesus monkeys are clearly at the latter end of the spectrum. They don't hesitate to punish the transgressions of those lower on the totem pole, which explains the victim's lack of resistance in this picture. It is better to sit still and oblige than to get bitten.

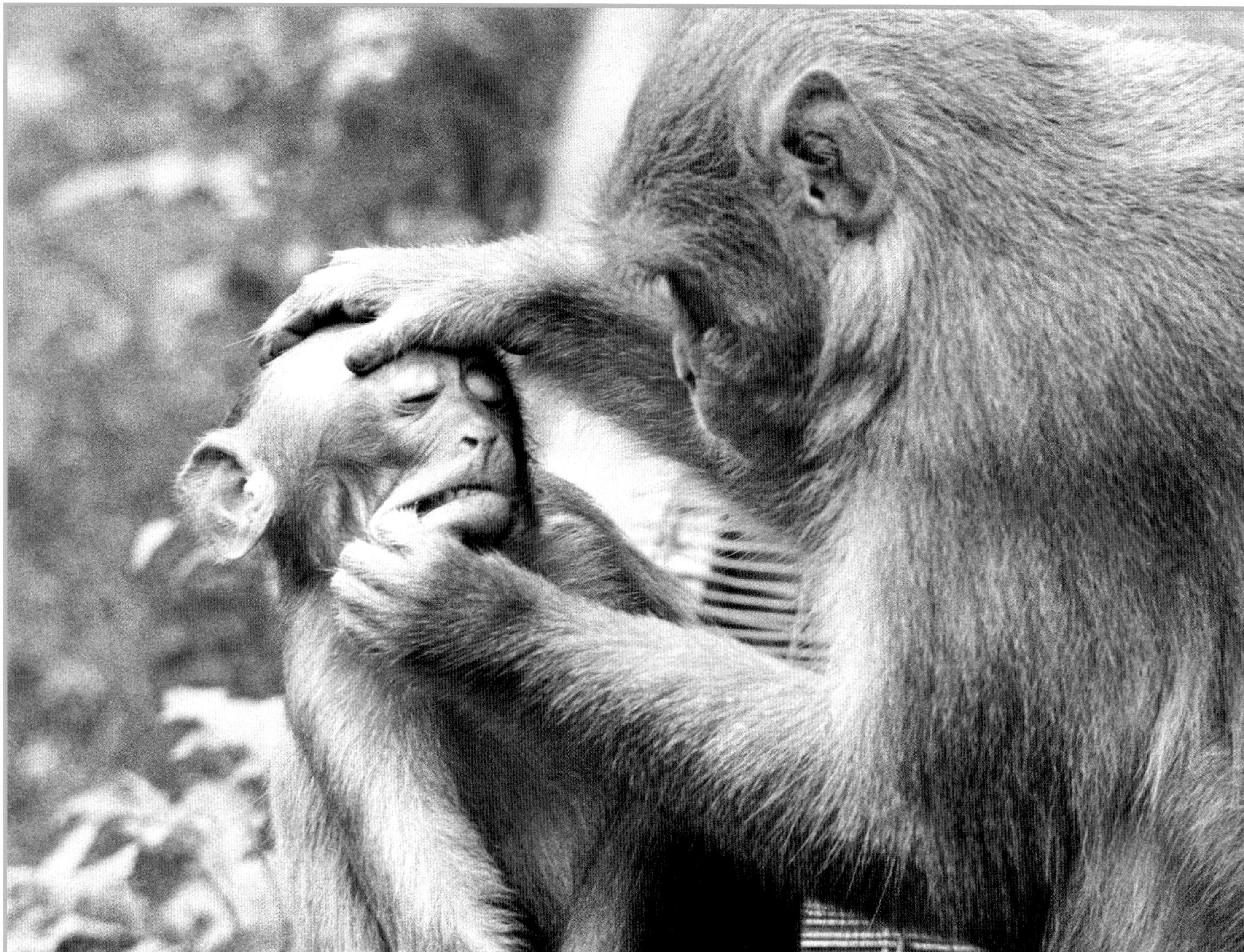

Give me back my berries!

Jakie was eating berries when a big male snatched them out of his hands. Instead of taking this lying down, he pursued the offender using screams and his species' typical gesture of begging with palm up. Jakie's vociferous protest contrasts with the rhesus monkey on page 95, who passively lets a dominant individual remove food from his mouth. Such is the difference between a relatively egalitarian, tolerant primate, such as the chimpanzee, and one who obeys a strict hierarchy, such as the rhesus. For chimpanzees, stealing is against the rules, and Jakie knows it!

Manual gesturing is uncommon in the animal kingdom: it is seen only in apes and humans. We immediately understand the meaning of Jakie's open hand, as it is also used by humans in need.

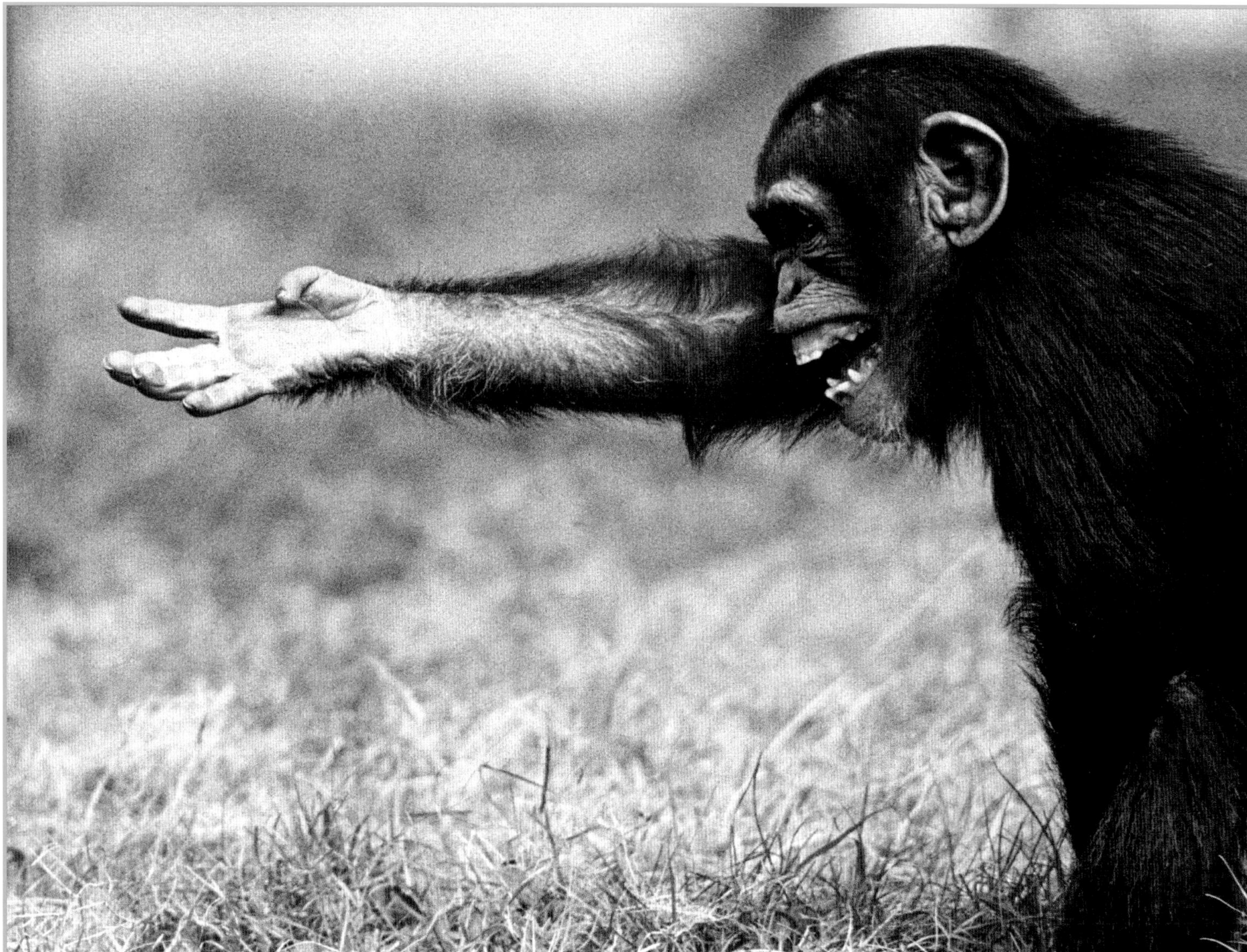

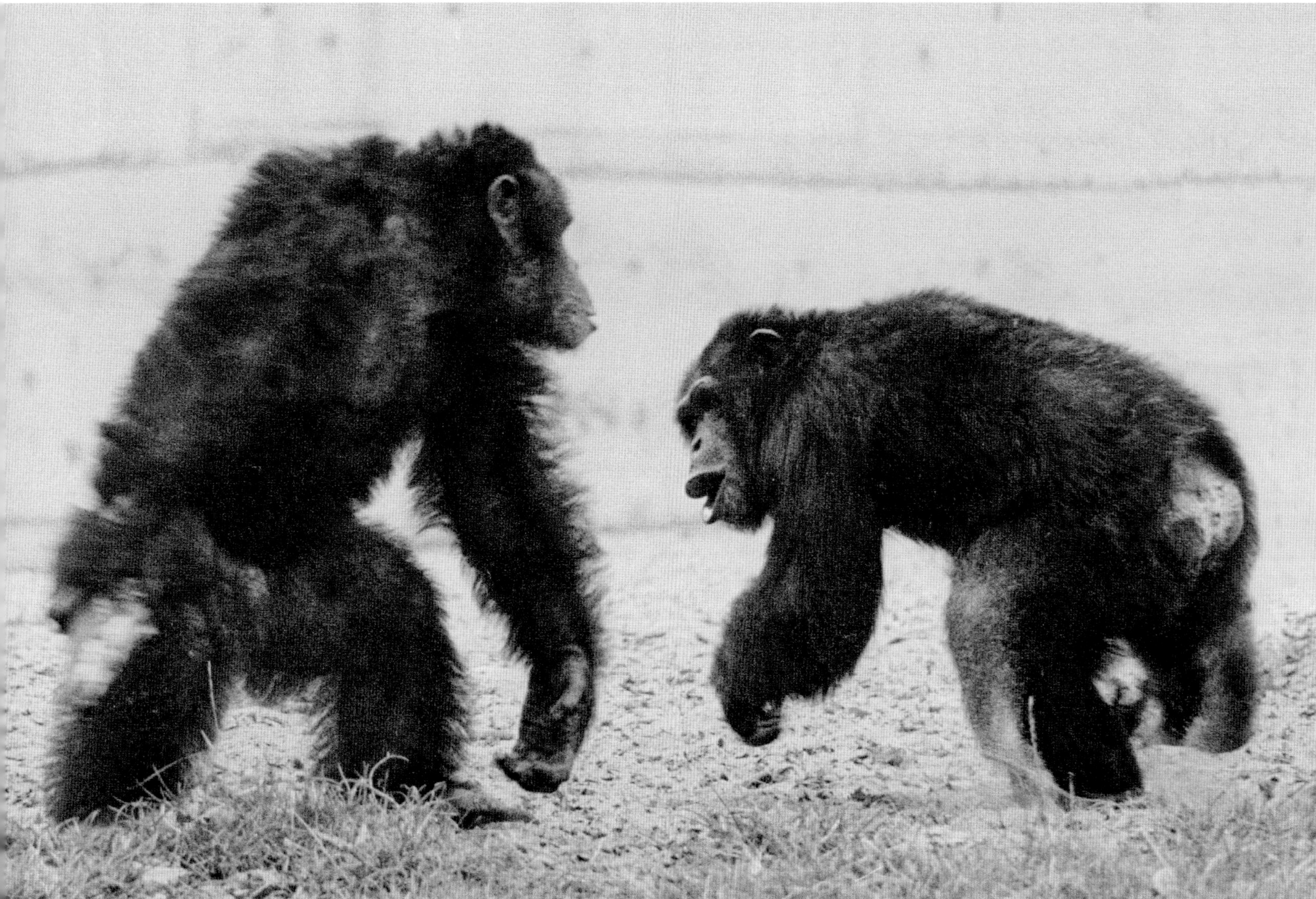

Big-man illusion

Chimpanzee males are constantly in the business of either emphasizing status differences or challenging them; seen here is a moment of confirmation in which Nikkie, the alpha male (left), brandishes a stick and makes himself look large by walking upright with his hair erect, whereas a subordinate male avoids him with pant-grunts, the chimpanzee's acknowledgment of inequality. Formalized rituals help preserve the peace in that they allow communication of the hierarchy, without risking aggression or open tensions. These rituals thus serve a function comparable to the rank insignia on military uniforms.

As alpha male, Nikkie had a custom of strutting around with heavy steps and hair slightly on end so that at first sight he seemed to occupy a position befitting his size and appearance (page 105). All of this was pure illusion, however, because Nikkie weighed no more than his rivals; he only *seemed* larger. He had adopted an appearance to fit his position.

Tit for tat

A chimpanzee tries to obtain food from another by sitting close and cupping her hand in a begging gesture that leaves no doubt about what she wants. Having noticed the mother of a newborn chewing food, the female on the left is interested (large photo). Even though she is of higher rank, the rules among chimpanzees, especially adults, require her to politely ask for a share. The response of the mother who is chewing may be to ignore her, walk away, or drop a few morsels into her held-out hand.

Active offering of food is rare, yet it does occasionally occur. In the small photo, a female "feeds" a juvenile as a reward for his long wait. Such behavior illustrates the capacity of chimpanzees to respond to the needs and wants of others. Sharing of food is subject to trade; that is, chimpanzees keep track of favors given and received and build an entire tit-for-tat economy of service exchange. From my studies, we know that food recipients pay back donors later by grooming them.

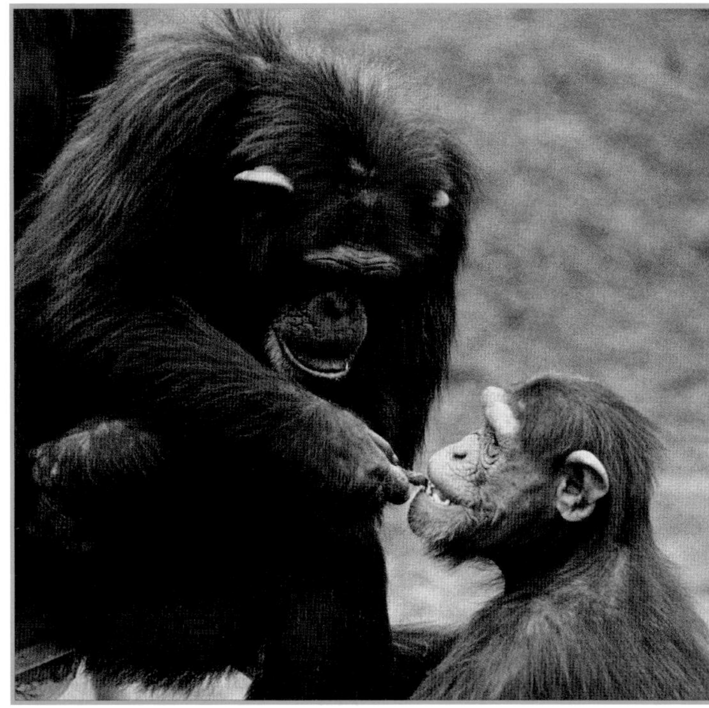

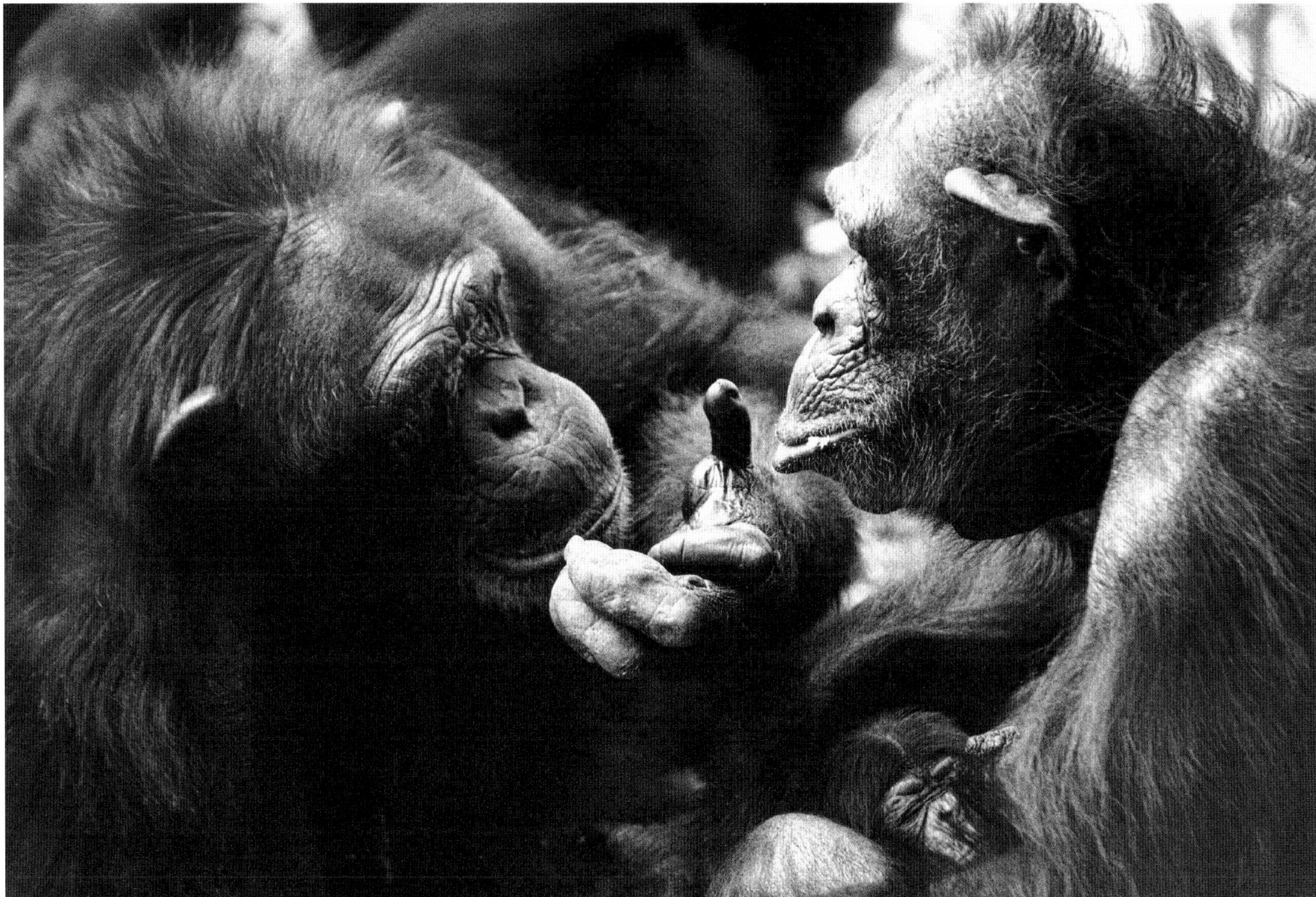

Cheek to cheek

A juvenile capuchin cups his hand next to the food an adult male is eating. He is about to press his cheek against the adult's, a gesture called cheek-to-cheek begging. Few primates are tolerant enough to let another touch their food, let alone take from it. When I first noticed that capuchin monkeys share food, I was puzzled. Except for mother-offspring sharing, other primates eat by themselves. But capuchin monkeys share just like chimpanzees and that other tolerant primate, the human.

A common explanation for food sharing among chimpanzees and humans is that it evolved along with cooperative hunting. Fruit and leaves are dispersed throughout the forest, as are insects and other small prey; there is no need to share small bites. But when several individuals join together to capture large prey, there must be rewards. Chimpanzees share meat after killing a monkey, and human hunters bring back prey to their villages. Recently, field-workers solved the capuchin puzzle when they reported that these monkeys, too, hunt in groups: they capture and eat large squirrels. This is a difficult task, requiring coordination among the hunters, a perfect setting for the evolution of sharing.

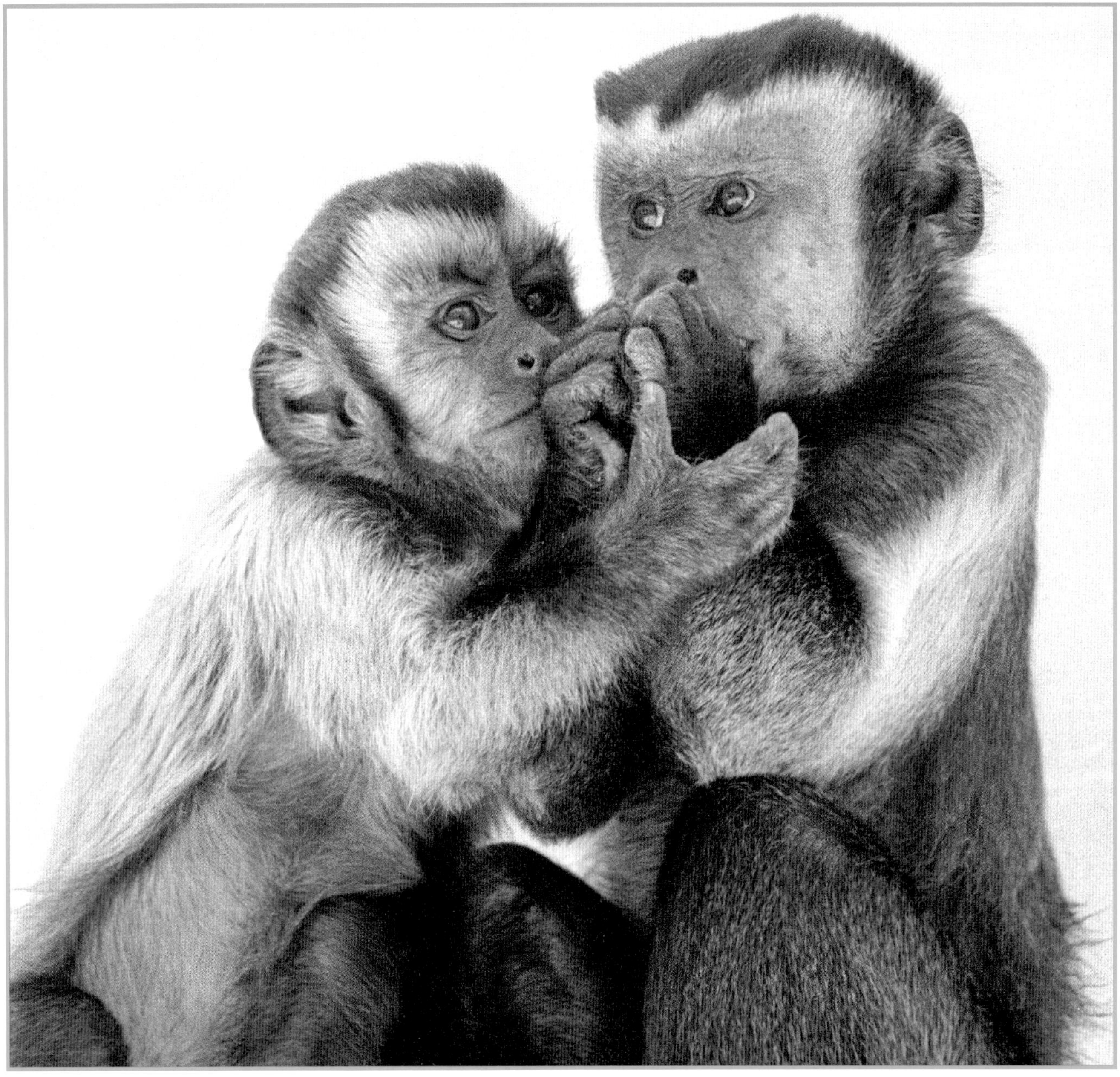

Brawn

Looking at Nikkie, seated here with his hair on end, one can see that a fully grown chimpanzee male is not to be messed with. Nikkie was only seventeen when this photo was taken but already rising in status in the Arnhem chimpanzee colony. With the support of an older male (pages 108–9), he had caused great turmoil by first rising in dominance over all females, then taking on the established alpha male. Muscular strength is only one factor in the success of a male chimpanzee, however. Strategic intelligence is another, and in this regard Nikkie was not nearly as cunning as his older partner, who used him as a pawn and broke off their coalition when other, more beneficial opportunities appeared on the horizon.

Chimpanzees are incredibly muscular. When scientists measured their pulling strength against that of American college football players, they found the apes to have five times more upper body strength than the healthy human males. This sets chimpanzees apart from the bonobo male, who is less driven to dominate and correspondingly much less of a bodybuilder type (pages 126–27).

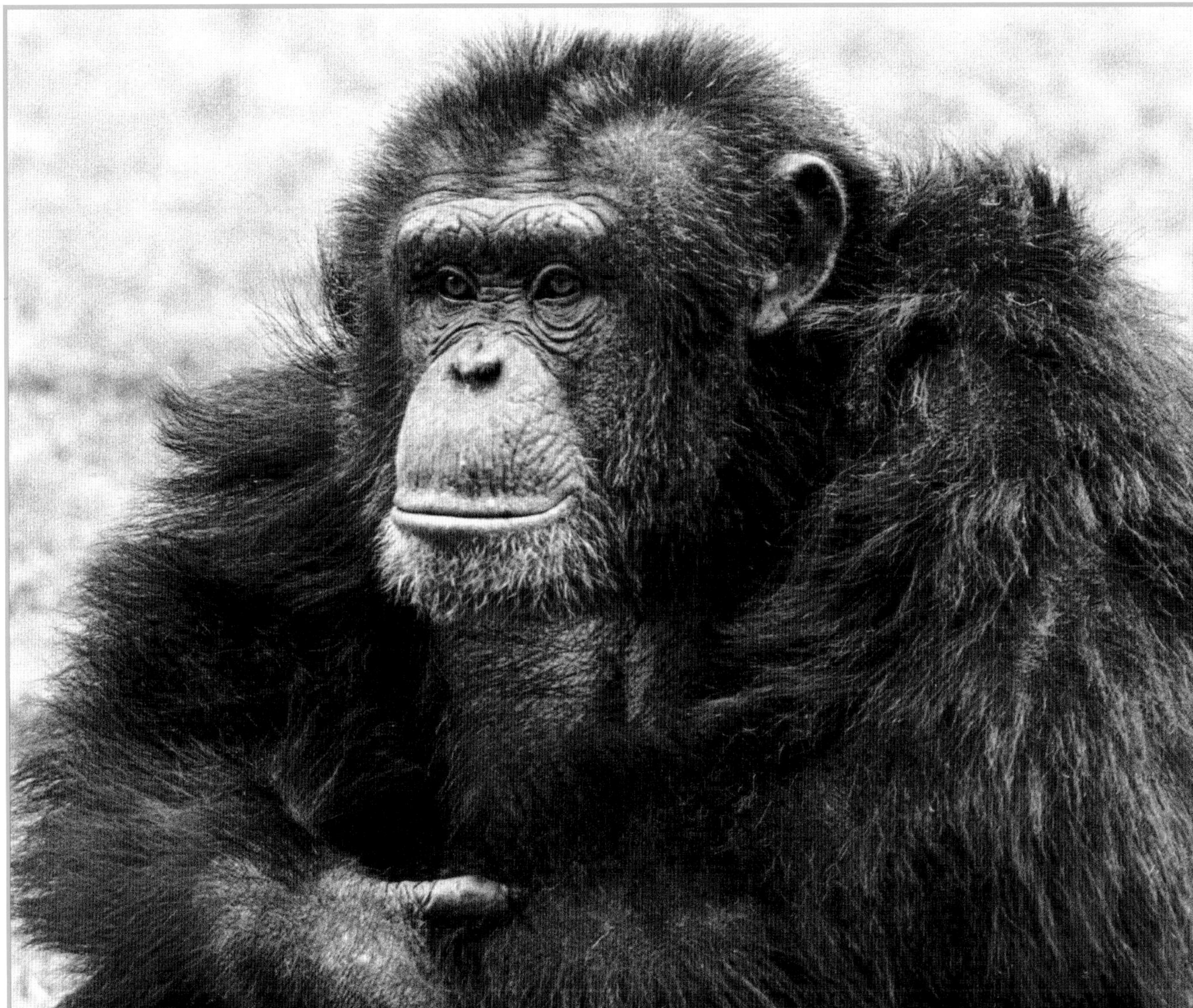

United we stand

If looks could kill, these two staring rhesus females would do a good job defeating their opponent. Standing shoulder to shoulder, mother (left) and daughter advance on another female who had threatened the daughter. Kin-based alliances explain why in macaque society a female inherits her mother's rank. Females receive so much support from their mothers, as well as from other female relatives, that they can be said to belong to a clique. If her clique is low ranking, a female will be treated like dirt. If her clique is high ranking, she will be treated like royalty.

Among chimpanzees similar coalitions can be found, but these are far more flexible and opportunistic, at least among the males. The two males in the small photo ruled the colony at the Arnhem Zoo for several years. Nikkie (behind) was the alpha and Yeroen (front) his older partner. Here they stand united, fiercely screaming, against their common rival. This partnership lasted only as long as Yeroen received sexual privileges from Nikkie for supporting him. When Nikkie started to become jealous and intolerant of Yeroen's contacts with females, their coalition dissolved, and Nikkie lost his elevated status overnight.

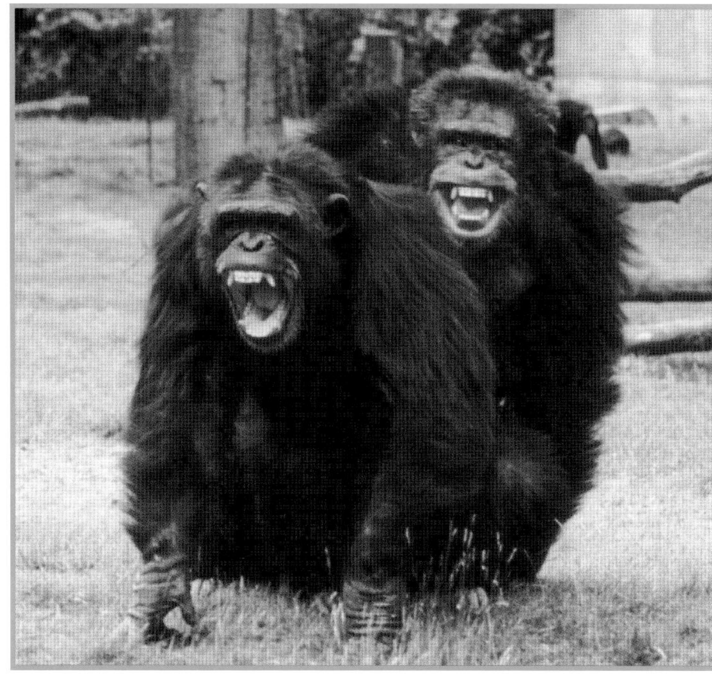

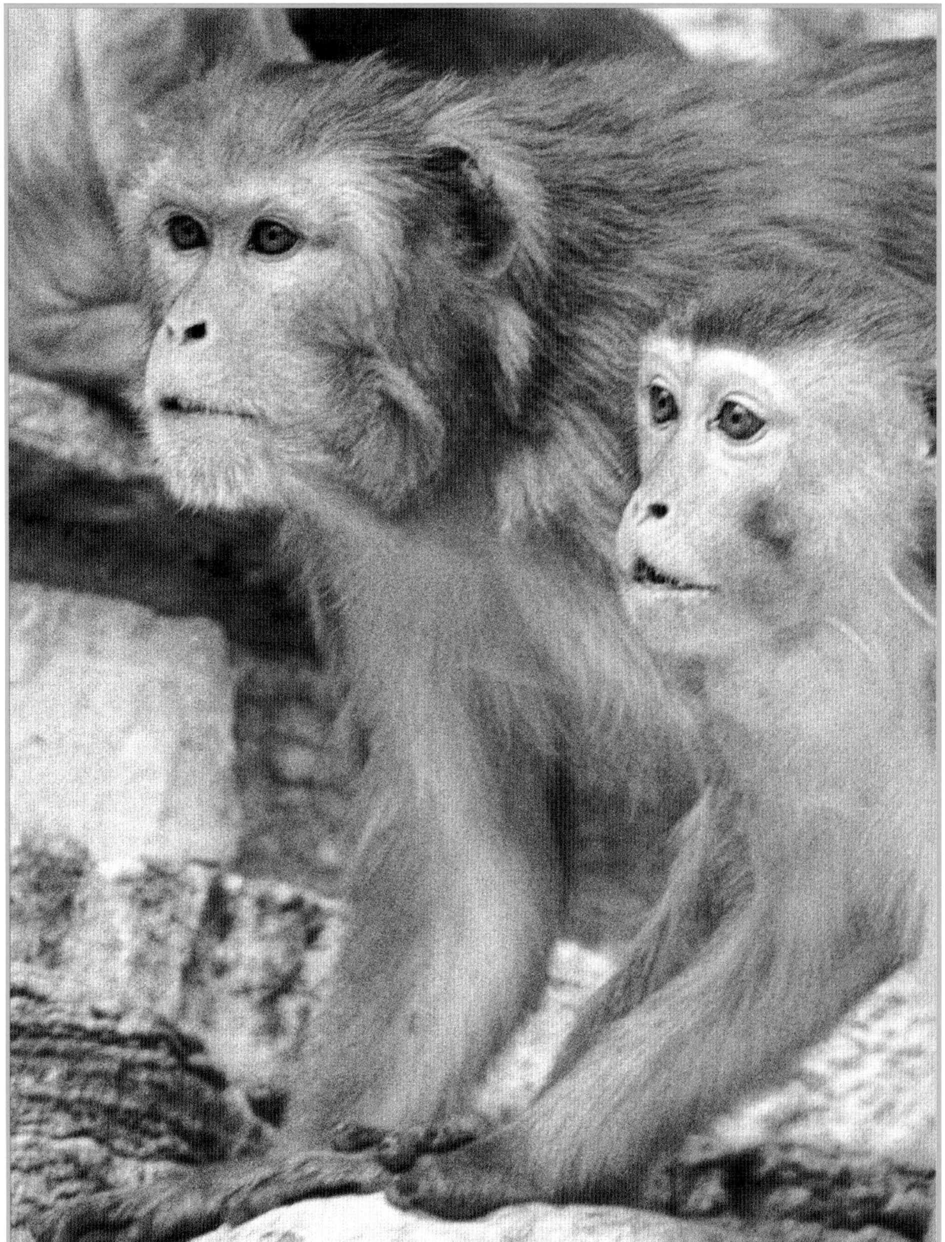

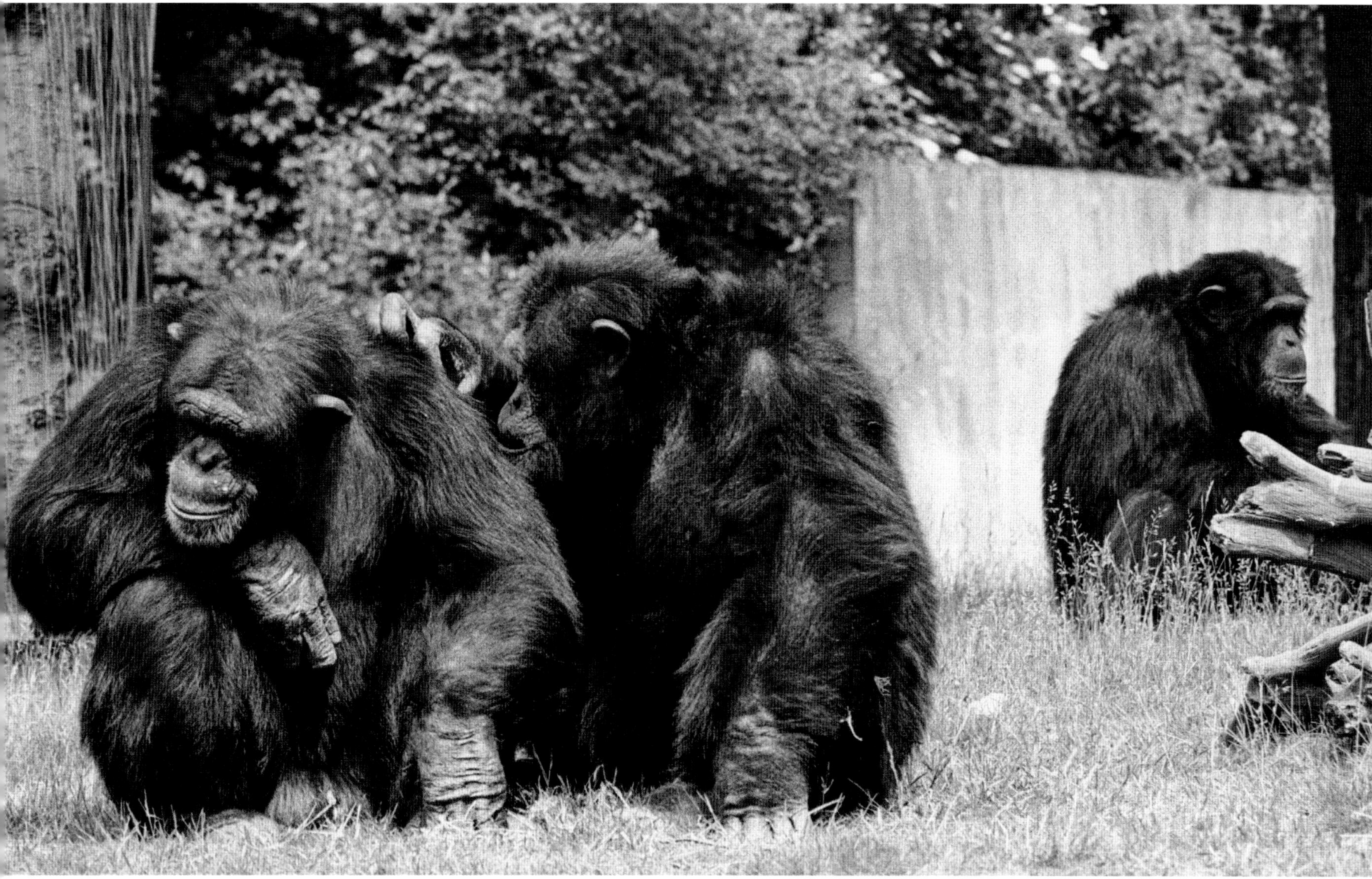

Chimpanzee politics

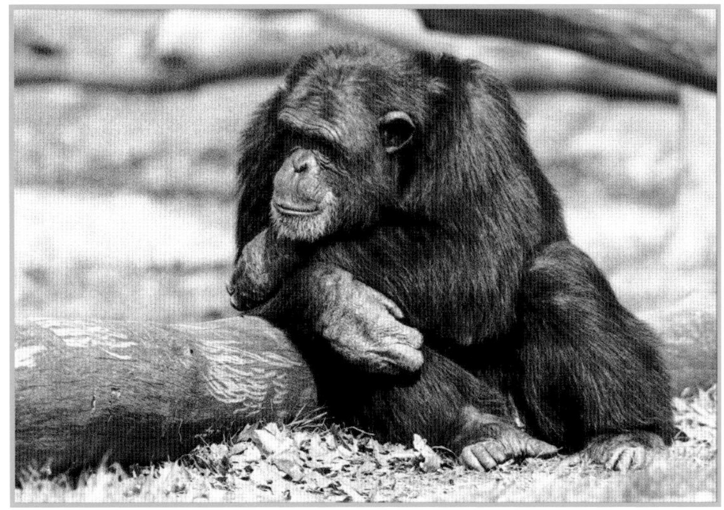

If chimpanzee politics can be summarized in a single frame, it is this one of three males at the Arnhem Zoo, two of whom have struck up a partnership. As in our own species, politics turns on who sides with whom and under which circumstances. Adult males are constantly jockeying for position. This leads to deal making such as that shown here: Nikkie (center) is grooming his older partner (Yeroen), while their common rival, Luit, sits looking rather unhappy in the background. Yeroen and Nikkie sometimes united against Luit (page 106).

But Yeroen, the kingmaker of this play, was not averse to occasionally also siding with Luit, thus keeping Nikkie on edge as well. In the wild, too, observers have noticed how elderly males play off younger males against each other. These senior statesmen make up for their lack of physical strength and stamina by exploiting the competitiveness of the young males around them. The small photo shows Yeroen resting, perhaps plotting a new scheme to increase his power.

Routing the enemy

One of the "natural" enemies of the rhesus monkey is *Canis familiaris*. In their native habitat in India, these monkeys hang around human settlements—sometimes they even live in large cities—as they have done for hundreds, perhaps thousands, of years. In this environment, dogs are the main predators. Monkeys have the advantage of being able to climb out of reach of canines. They are also excellent at banding together, as occurs here with two adult males. Together the monkeys can tackle an enemy several times their size by surrounding it and threatening it from different sides. I saw one monkey pull the dog's tail while another stayed in front of him to hold his attention. Rather than an Indian setting, though, this photo shows a farm in Wisconsin where a laboratory group of rhesus monkeys had been released to see how they would adjust to an open, green environment (which they obviously had no trouble with). This farm had only one dog, which put the monkeys at a distinct advantage. In India, by contrast, they often face packs of canines, a species as adept at coordinated action as monkeys.

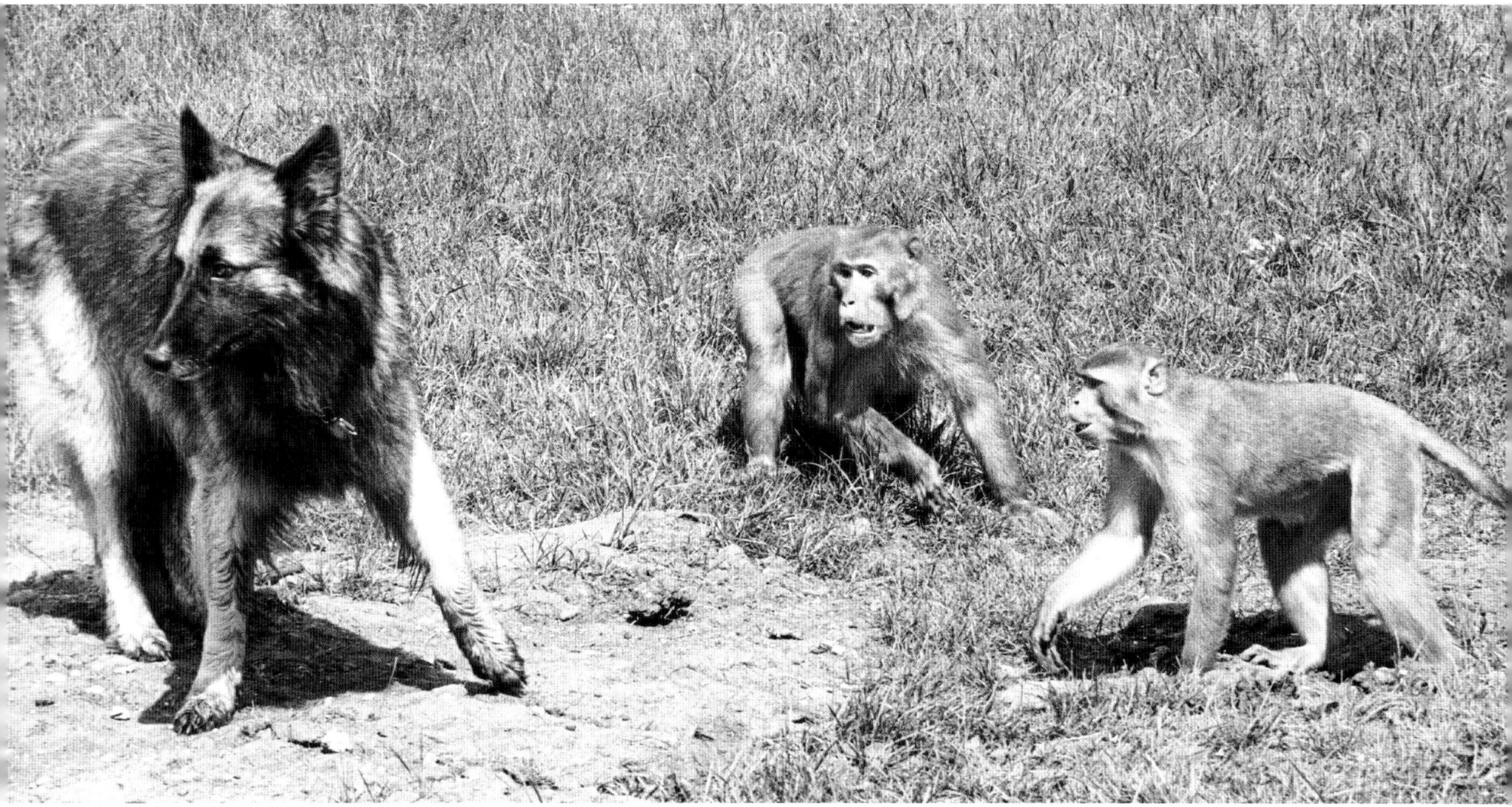

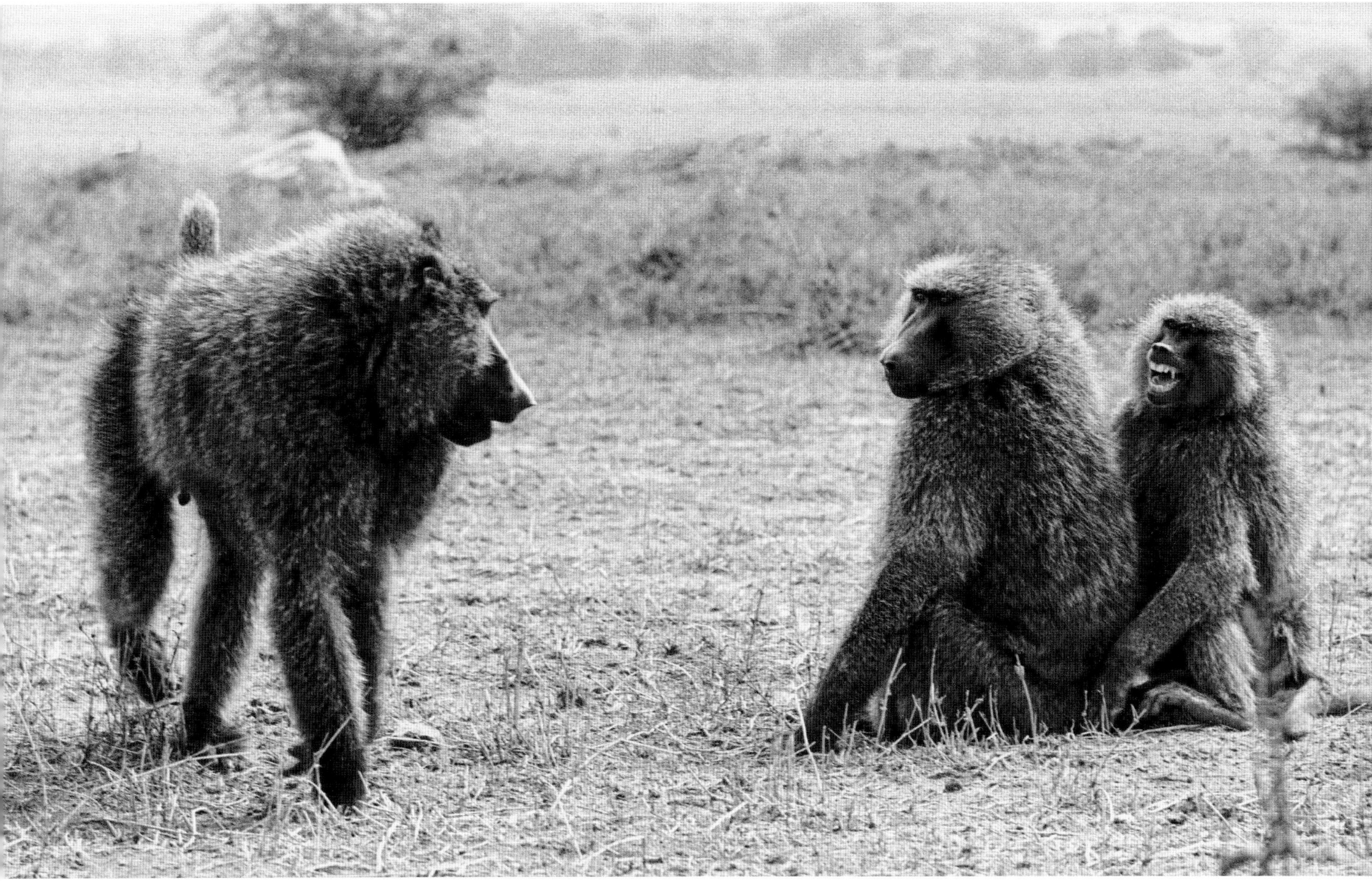

What are friends for?

Residents of a baboon troop will resist the entrance of a new male. The males are unhappy to see a new rival on the horizon, whereas the females dread being harassed by the newcomer. Females usually have male friends to defend them against strange males, which is a necessity given that male baboons are twice the size of females and armed with formidable teeth. The female on the right has moved out of the way of an aggressive new young male on the left. She clings to the back of a resident male, who protects her by staring down the other male. Such special relations between male and female baboons have been dubbed "friendships." Friendships are expressed by the baboons' sleeping together, grooming of a male by a female, and protection of a female and her offspring by a male.

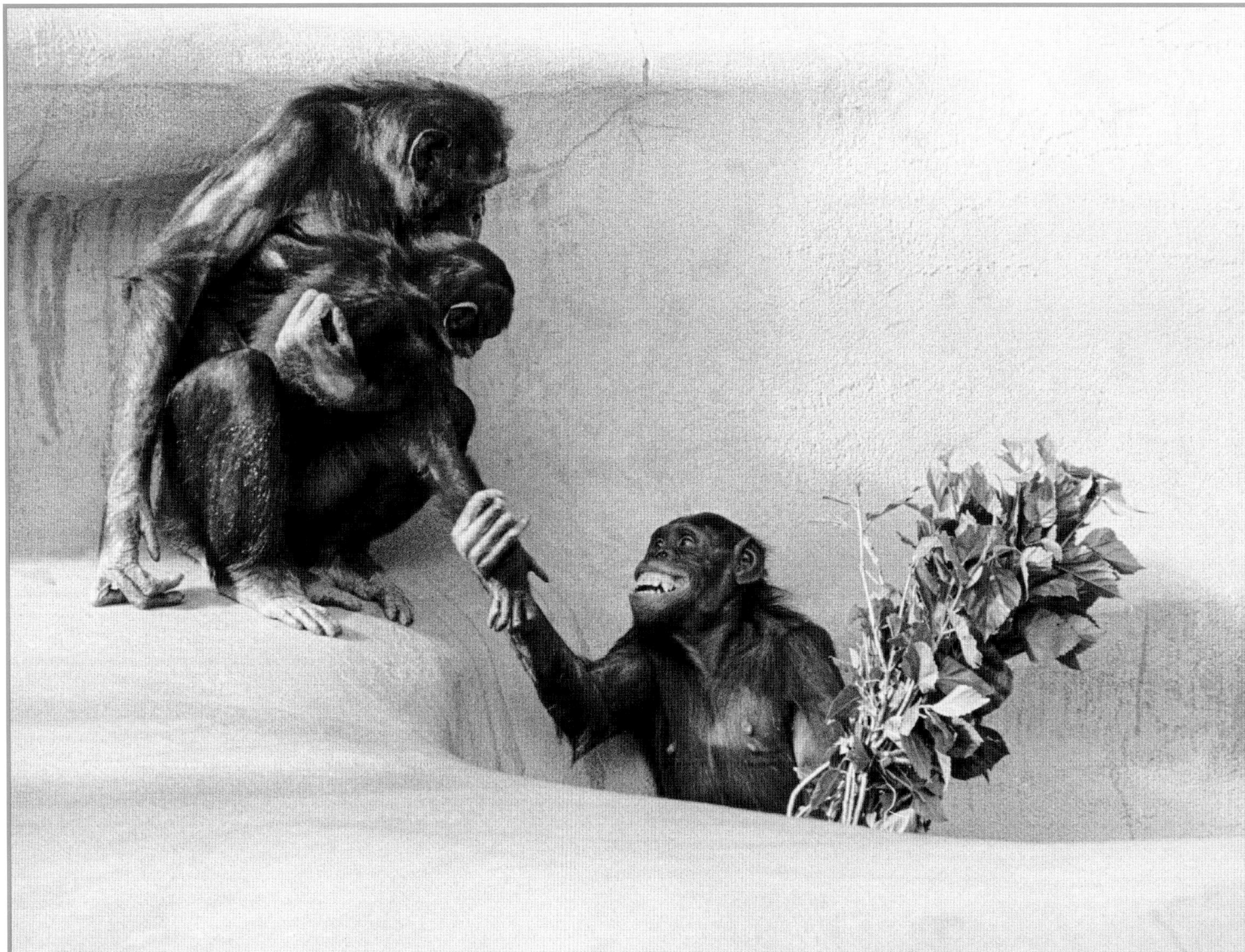

A handshake and a smile

Loretta (right) holds an attractive bundle of hibiscus leaves provided by the caretakers at the San Diego Zoo. She has moved her prized possession out of reach of Lenore, an infant in her mother's arms, who had made a brazen grab for the food. An adult would not make such an attempt, but Lenore is only three years old. Like all youngsters, she knows how to take advantage of the close presence of her mother, Louise, who dominates Loretta. This poses a tricky problem for Loretta: she needs to resolve the skirmish without upsetting the infant, since Louise is very protective. The big grin on Loretta's face signals appeasement. The grin is directed at the infant, whose hand Loretta holds in an attempt to calm her down so that she herself may eat in peace. She will not totally monopolize the food, though—once she has broken the bundle apart, there will be shares for the others.

Telltale forehead

In the 1980s I missed working with chimpanzees so much that I decided to try capuchins, a Neotropical primate also known as the organ-grinder monkey. The capuchin is the "South American chimpanzee" because it shares many characteristics with its more famous distant relative. Look at this male's high forehead, for instance. Among the primates, capuchins possess one of the largest brains relative to body size. They also live a long time (sometimes more than fifty years in captivity), establish male political coalitions, excel at using tools, and share food (page 103). All of this hints at convergent evolution between capuchins and the great apes. However, since the New World primates, to which the capuchins belong, split from the Old World primates at least thirty-five million years ago, the similarities evolved independently on different continents.

Of course, major differences exist between capuchins and apes. The capuchin is small, only the size of a cat. In terms of intelligence, the capuchin may be remarkable, yet it is not in the same league as the great apes. And its prehensile tail is a wonderfully flexible extra limb, visible in a jumping wild capuchin.

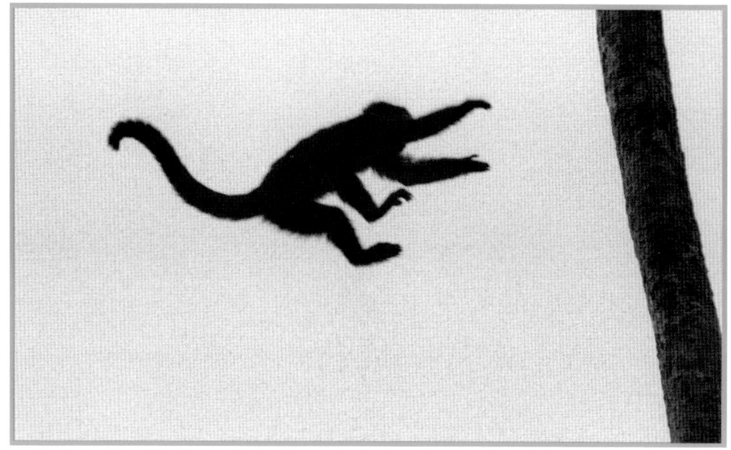

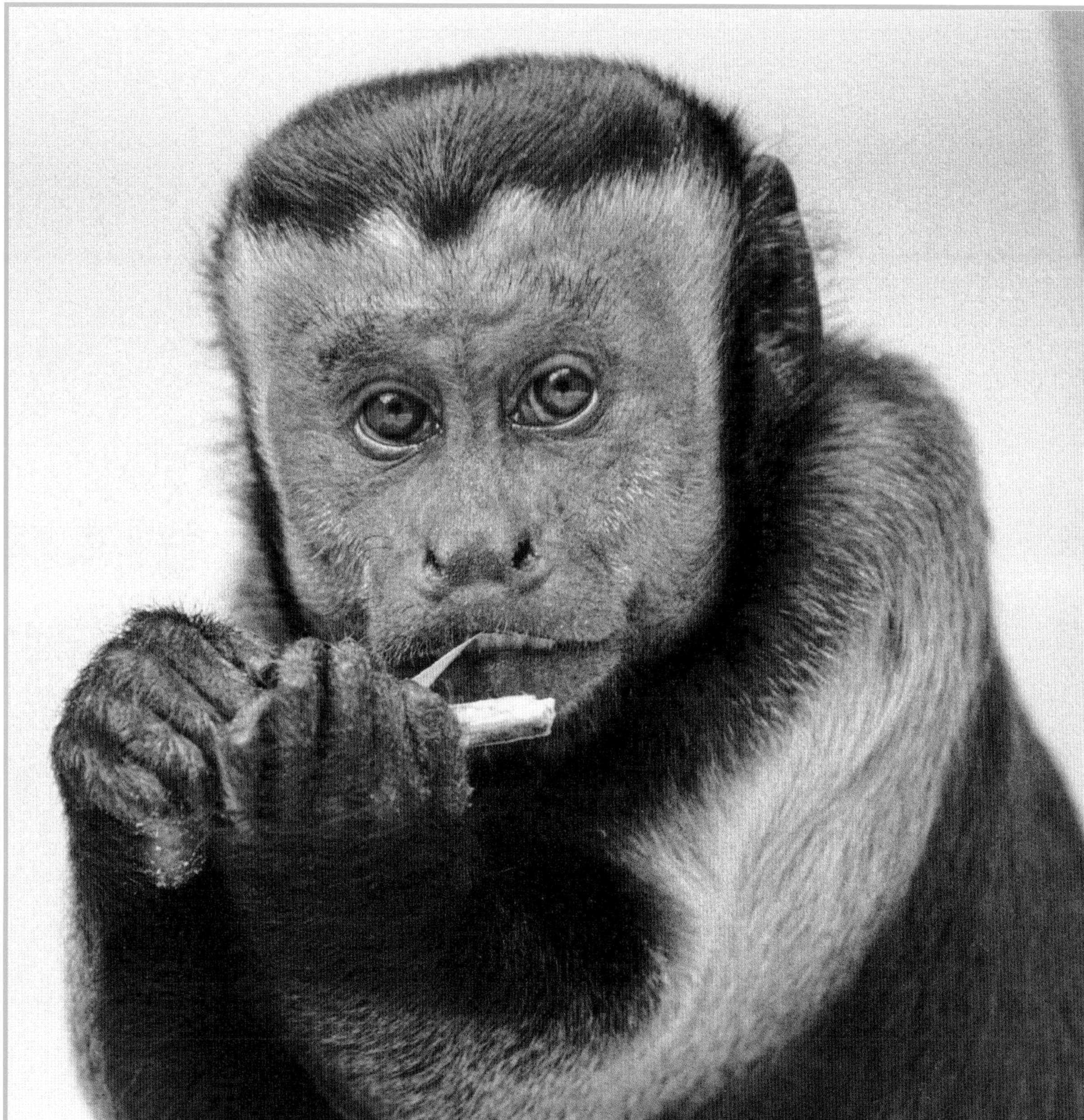

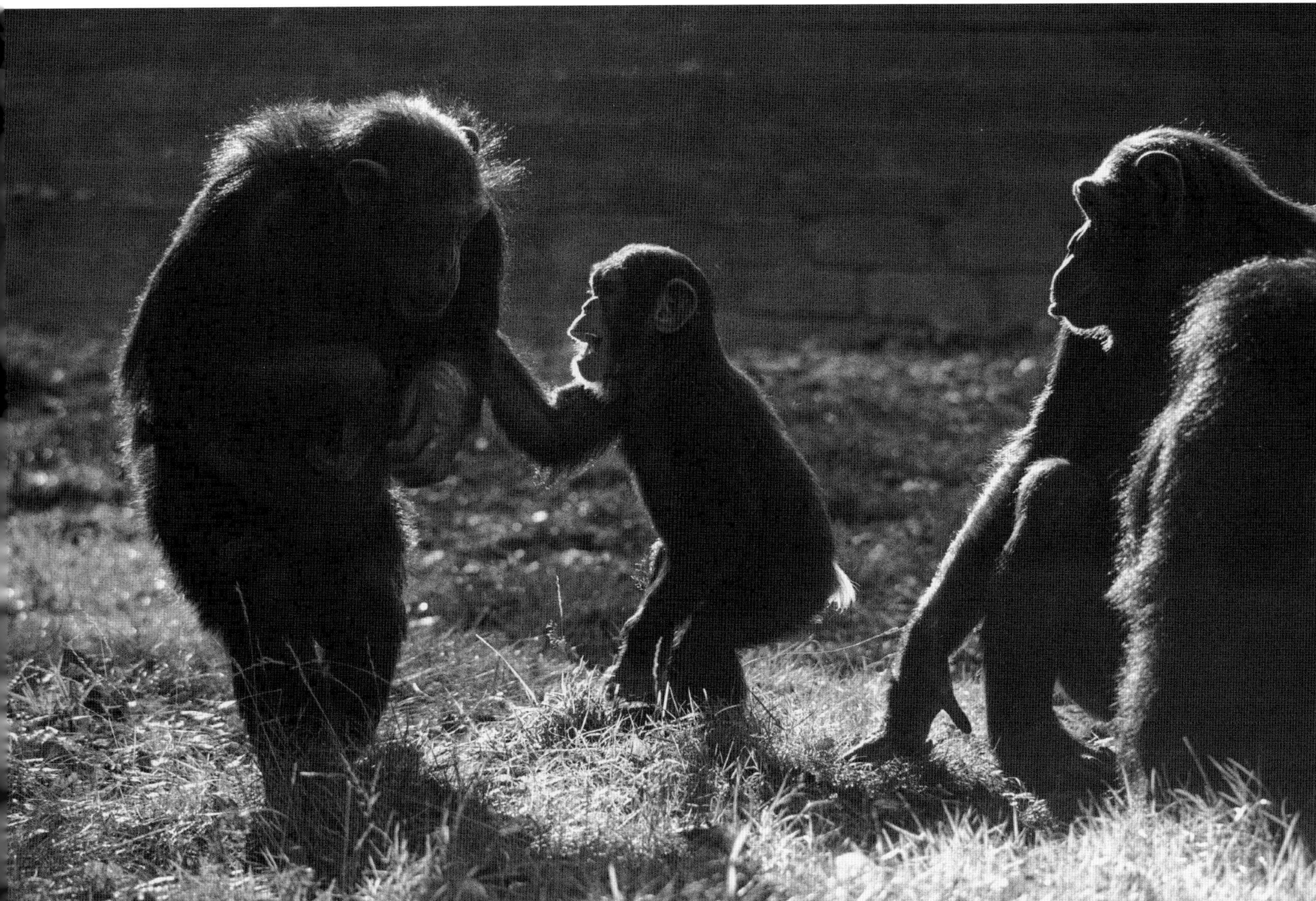

Morning fog

Chimpanzees hate getting wet. During or after a downpour they often walk on two legs, as if they wish to keep at least their hands dry. Here Zwart (left) joins a group in the morning fog at the Arnhem Zoo, walking hunched up with her hands folded in front of her because of the dew on the grass. I have never understood this aversion to wet hands, but then again humans walk bipedally.

The Arnhem colony, the largest in the world, has spontaneously split into subgroups of apes who hang out together. Each individual belongs to a small circle of friends and family with which he or she feels most comfortable. Amber, with the sun-outlined facial silhouette on the right, is the leader of a subgroup of young adult females. She is also a caring auntie to little Moniek, the daughter of the colony's powerful alpha female (pages 34–35). Moniek therefore spends much time with this subgroup. She (center) welcomes Zwart with an open-mouthed play face and a jovial slap on the back, as if commenting on her miserable posture: "Come on, it's not *that* bad." The sun is coming through, and soon everyone will be drying out.

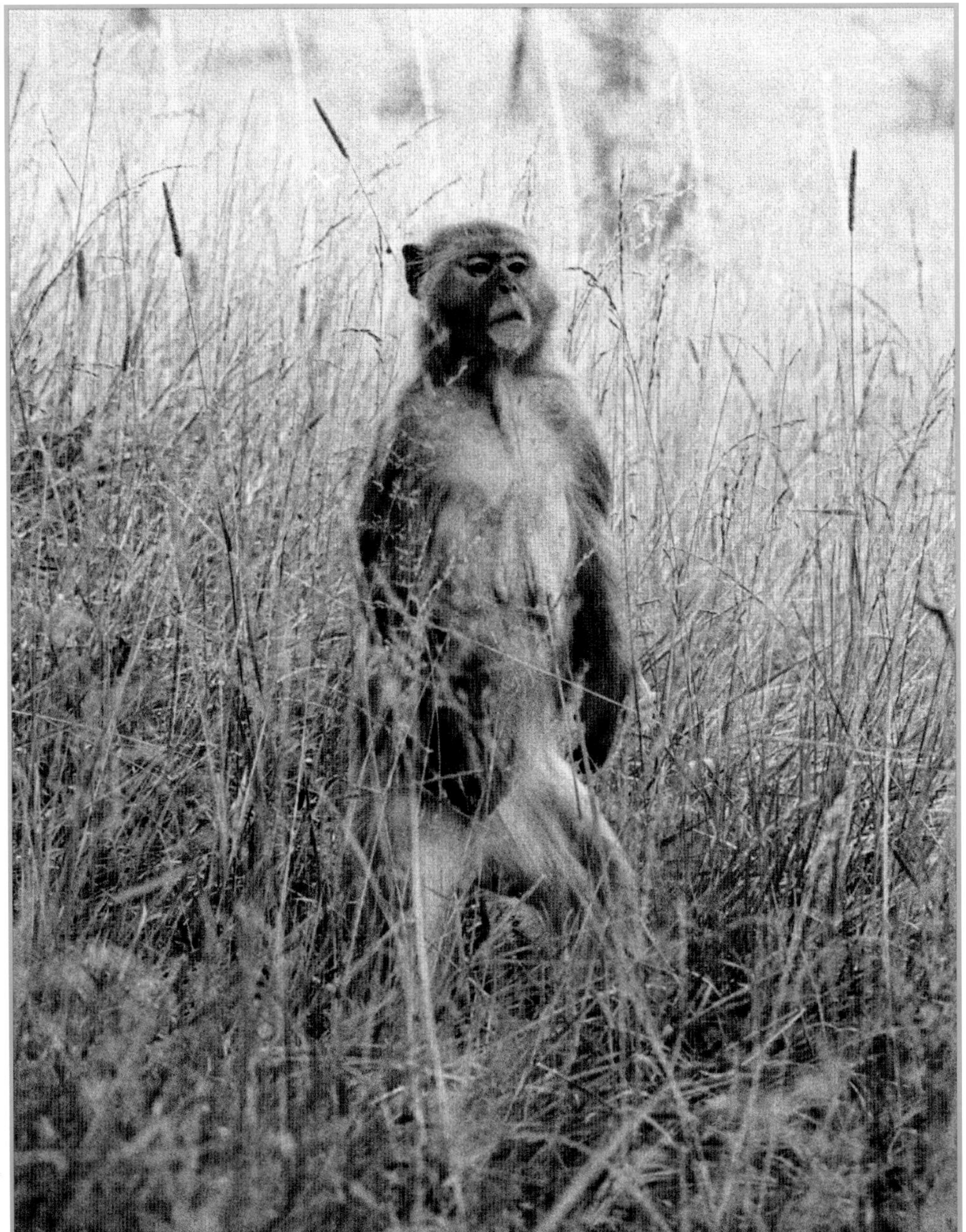

Standing tall

Why are humans the only primates to habitually walk on two legs? Theories abound, and a major one is that when our ancestors ventured onto the savanna, it became useful for them to look out over the tall grass in order to spot predators and keep an eye on other members of their traveling party. As if to illustrate this point, a female rhesus monkey stands upright in a meadow (large photo). She does so only briefly, though, and without walking.

The trouble with the lookout theory is that being taller than the grass also makes it easier for predators to spot the individual who stands up. Possibly, then, there were serious costs associated with bipedalism. Alternative theories assert that protohominids began walking on two legs in order to free their hands to carry food and weapons, or that they protected themselves against ultraviolet rays by reducing the amount of body surface exposed to the sun. It has also been suggested that our ancestors drew attention to communicative gestures and facial expressions by standing up, exposing a frontal body view. This display theory is illustrated by the female chimpanzee in the small photograph, who is screaming and hitting herself with her arms in frustration about a rejected request for food.

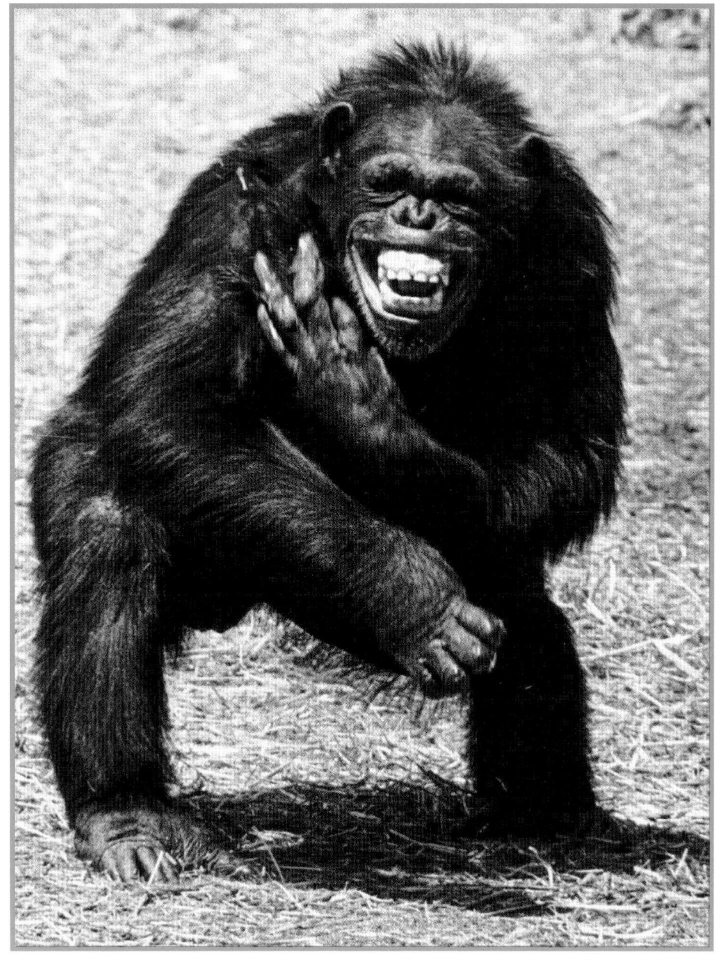

Like Australopithecus

With respect to body proportions, the bonobo looks eerily like an artist's impression of Lucy, the well-known *Australopithecus afarensis* fossil. The bonobo's long legs and straight back when standing upright are quite different from the stumpy, buckling legs and sloping back of the other apes. Bonobos look extremely comfortable standing up and walk with impressively self-assured strides.

Nevertheless, bonobos are anatomically far from little people. Two major differences in general anatomy can be seen right away in this shot of an adult female (left) and adolescent male on the lookout for the zookeepers with their meals. First, the bonobo's foot, with its opposable big toe, is evidently that of an arboreal primate: it is designed to grasp and hold on to branches. Bonobos are first of all climbers. The second difference is the length of the bonobo's arms: no person I know can reach with a hand below his or her knees while standing up straight. Yet, apart from these striking differences, the bonobo's anatomy is the most anthropoid (humanlike) of the apes.

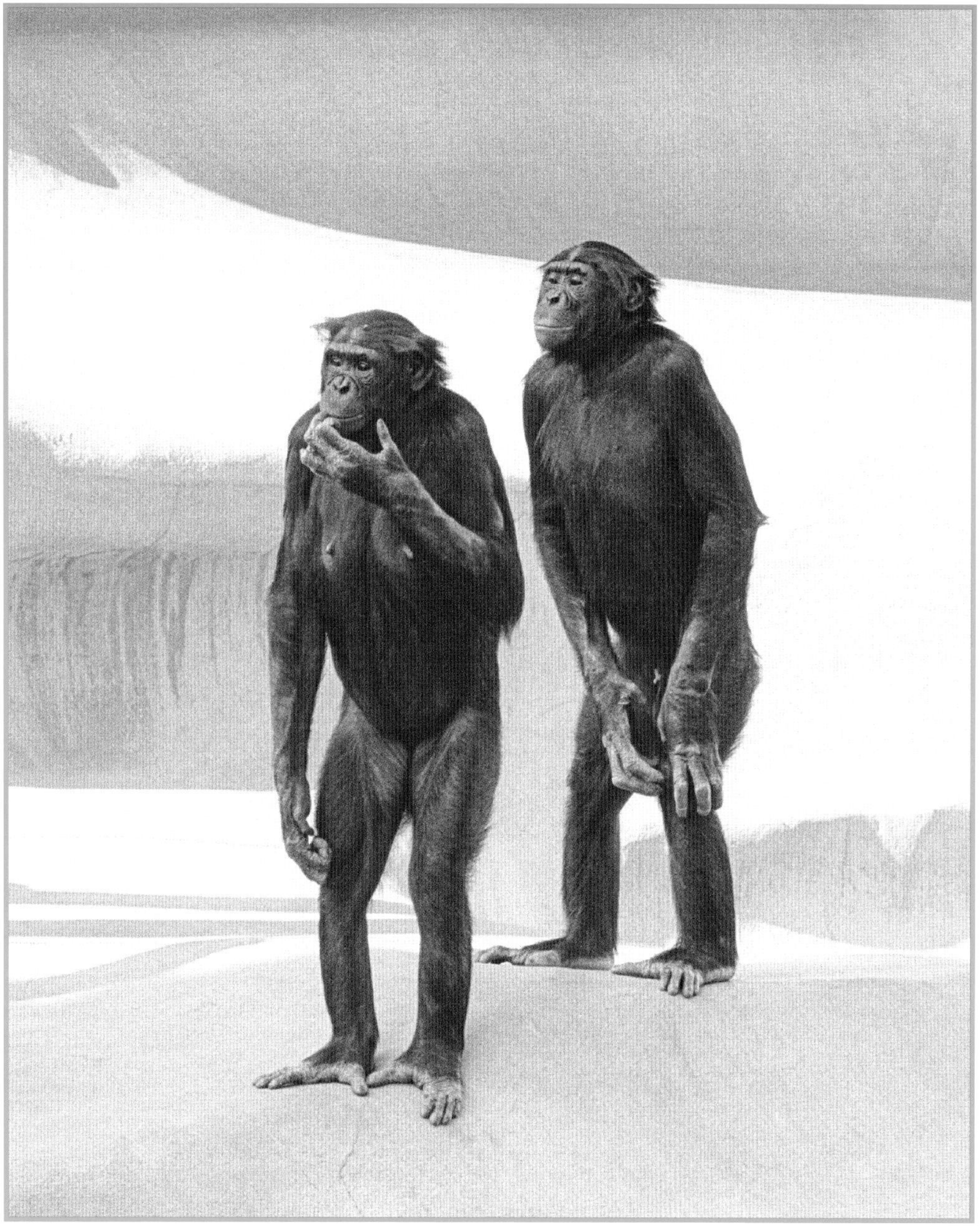

Ropewalker

As expected of an arboreal creature, bonobos have a perfect sense of balance. This adult female doesn't even need to concentrate on her locomotion: she calmly munches on leaves while rope-walking from A to B.

The way bonobos relate to and use their bodies can strike humans as unusual. For example, their feet serve as hands. They grasp things with their feet, gesture with their feet during communication, and clap them together to attract attention. Apes are sometimes called quadrupedal (four-footed), but a better term for bonobos might be *quadrumanual* (four-handed). They are more acrobatic than any ape, jumping, brachiating, and flipping around with unbelievable agility and sureness of movement—not to mention the myriad ways in which they achieve copulation, which challenges the *Kama Sutra*.

The bonobo's acrobatic talents relate to the fact that they have never been forced, not even partially, out of the forest. They still live in the same humid forests believed to have been the habitat of the common ancestor of humans and apes. In other words, bonobos never had to abandon or compromise their tree-dwelling ways.

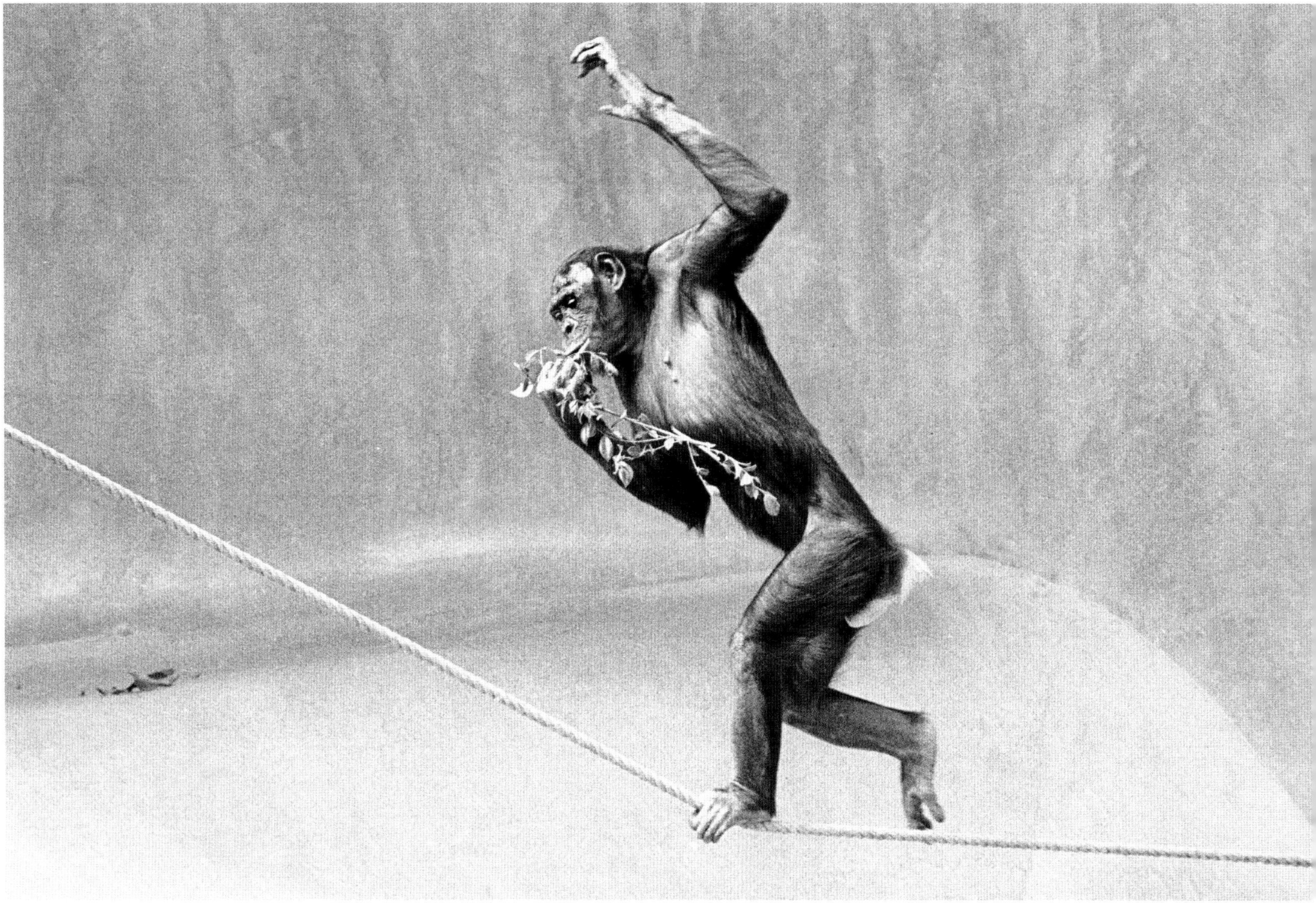

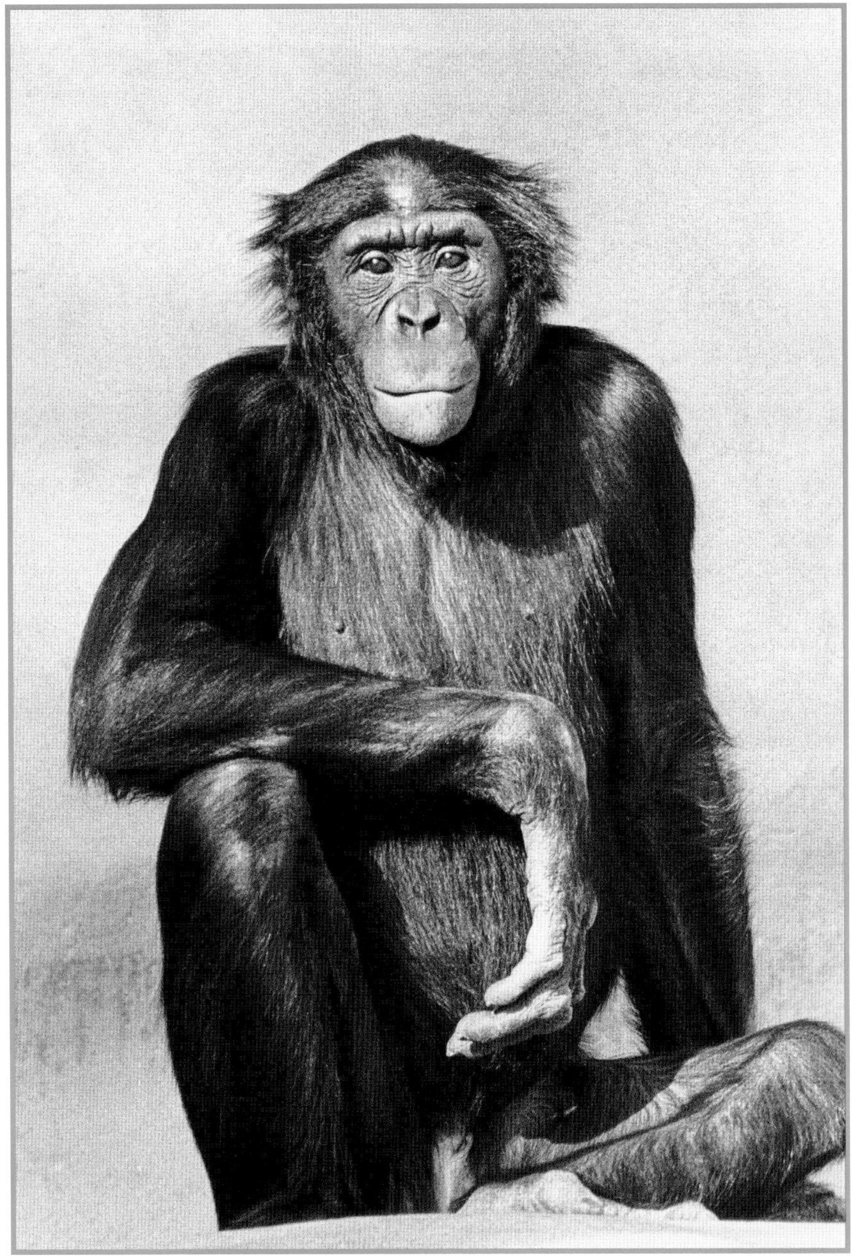

The gentleman

Kevin is a real gentleman: always considerate of others, never mean or demanding, and generous and gentle. He exemplifies the delicate features that make the bonobo so elegant compared to the other apes: the special hairstyle with a part in the middle, slight build and narrow shoulders, long arms, and piano player hands.

I don't want to insult any chimpanzees, but compared to bonobos they look coarse and unintellectual. The portrait of chimpanzee Nikkie (page 105), makes it obvious which of the two species packs more muscle and which is more pacific and sensual. As the first description of live bonobos put it, translated from German: "The bonobo is an extraordinarily sensitive, gentle creature, far removed from the demoniacal *Urkraft* [primitive force] of the adult chimpanzee."

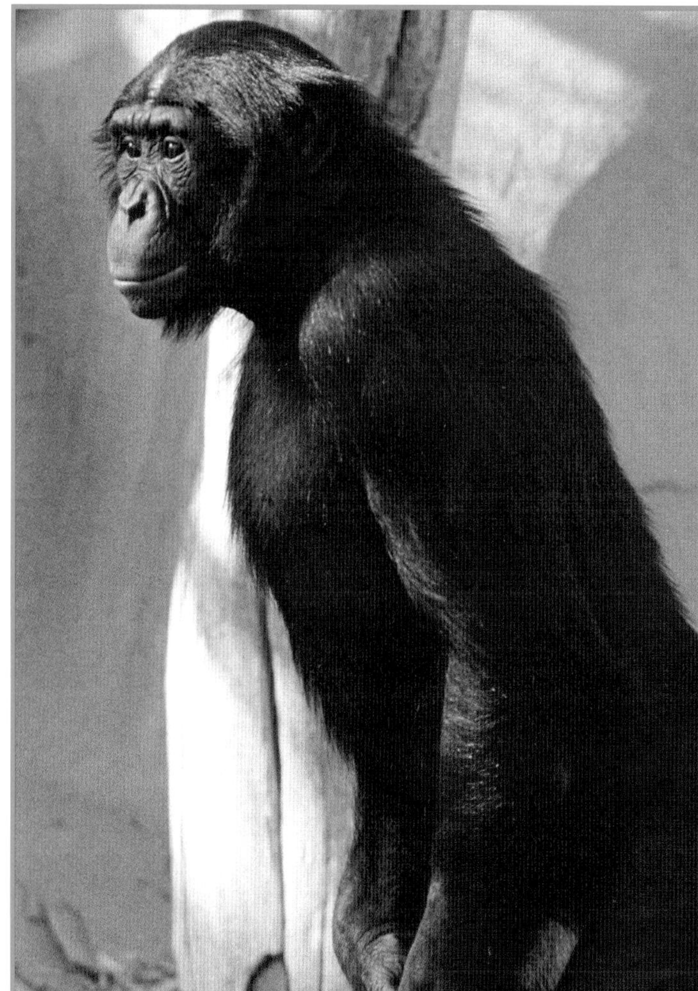

Survival of the unfittest

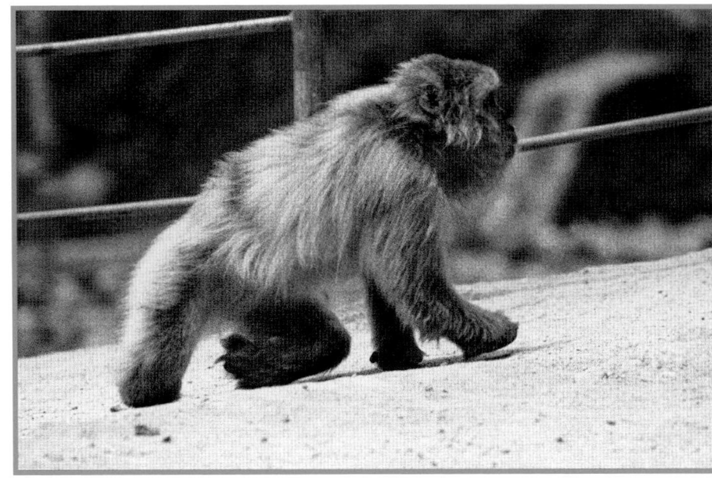

Mozu's face, photographed when she was nineteen, betrays lifelong suffering. Mozu is famous in Japan, having starred in several television documentaries. Born without hands and feet, she had trouble keeping up with the rest of her troop in Jigokudani Park in the Japanese Alps, near Nagano. This was especially true in the winter: when the other monkeys traveled through the tree canopy, Mozu had to plow through the cold snow on the ground. Nevertheless, she survived the harsh climate and successfully raised several offspring.

This monkey is living proof that "survival of the fittest" doesn't always work the way people think. In highly social animals, such as macaques and also dolphins, wolves, elephants, and people, the individual is a member of a group, and the group does not necessarily reject or expel handicapped or sick individuals. These individuals sometimes enjoy extraordinary tolerance. Mozu was a well-accepted member of her society, which goes a long way toward explaining her survival.

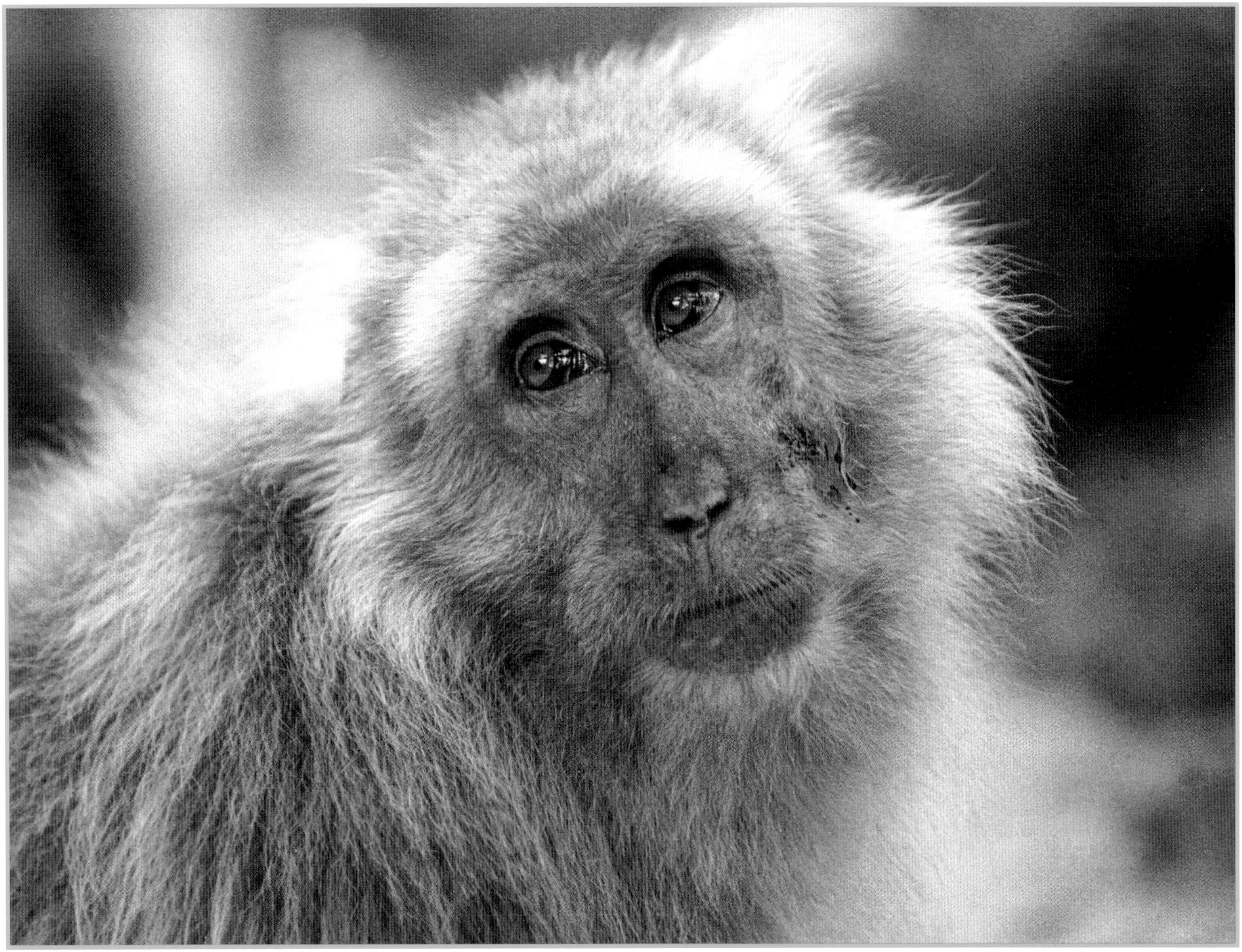

See no evil

The Japanese have a long history of worshipping wild animals and provisioning them, such as the famous deer that for more than a thousand years have roamed the temple grounds around Nara. The presence of native primates has strengthened the sense of connection people feel with nature, a sense often conspicuously absent from Western thought. Eastern folktales and poetry are laced with references to monkeys. The three wise men, or Magi, of the Bible are paralleled in the East by the three wise macaques of Tendai Buddhism ("See no Evil, Hear no Evil, Speak no Evil").

At Katsuyama, not far from Osaka, I encountered a group of monkeys in a small park with a waterfall. Tourists would come to feed the wild monkeys in the way that we feed ducks in the pond. Even though macaques are small by human standards, and clearly are quite different from us bipedal apes, their expressive faces leave no doubt about the continuity. When this eight-year-old male briefly glanced back over his shoulder while sitting against the sun, I felt a jolt of recognition.

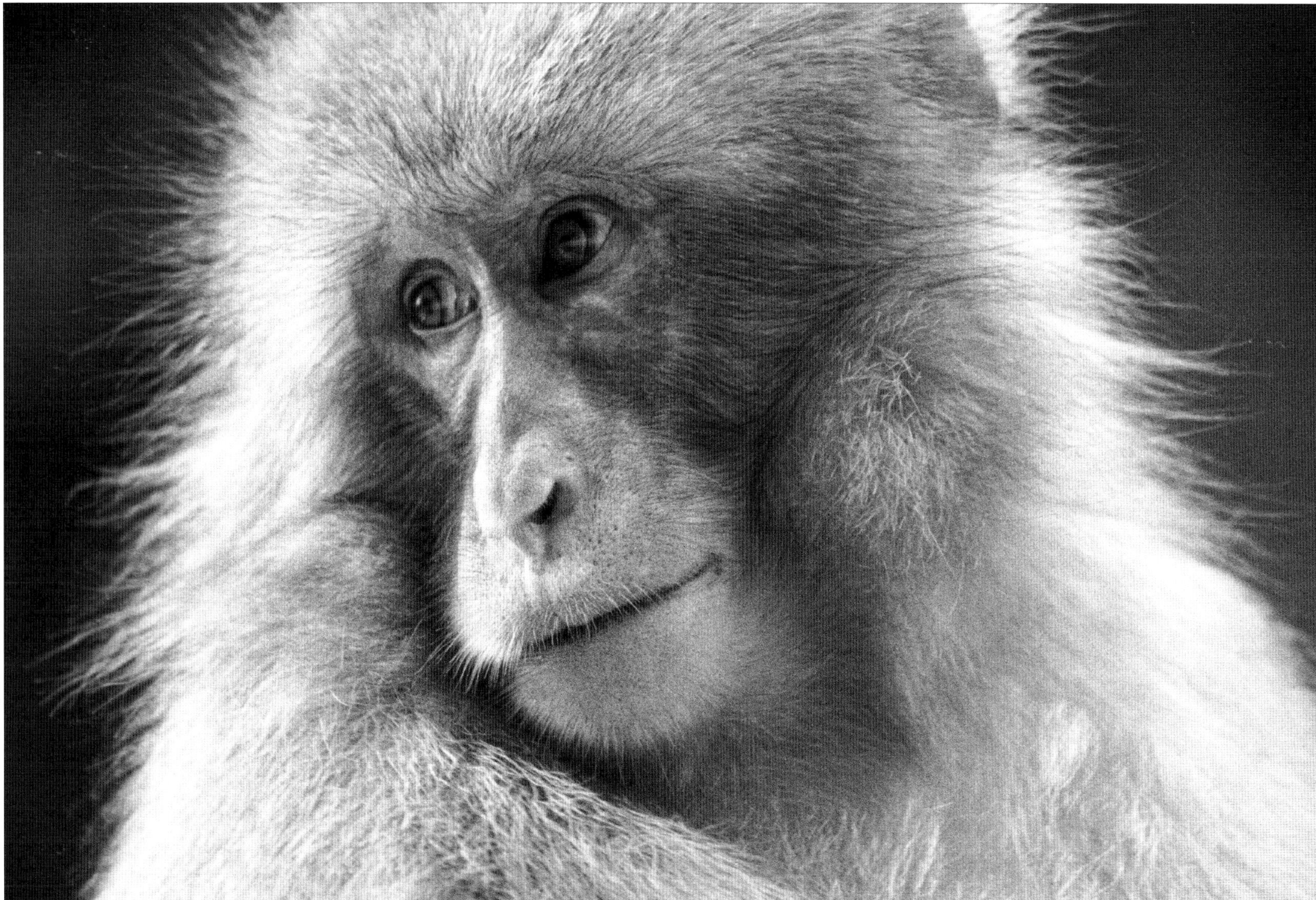

The appearance of thinking

We love to present the human philosopher the way Auguste Rodin did in *The Thinker*, with his chin leaning heavily on a fist. Kevin, an adolescent male bonobo, seems to be thinking equally hard, perhaps about what distinguishes bonobos from other animals. I could help him with this, because it is obvious that the species is a lot sexier than most animals—with the possible exception of their close relatives *Homo sapiens*.

Even though this picture conveys a reflective mood, and actually graces the cover of a book on the origins of intelligence, I must admit that the reality is more mundane. When I studied the bonobos at the San Diego Zoo, their enclosure was right under the path of cable cars, which offered visitors a nice bird's-eye view of the grounds. Some visitors could not resist breaking the rules by feeding animals below. A few candies had just landed in the enclosure. The apes had quickly gathered them, and here Kevin is looking up at the cable car to see if more is forthcoming. Thus, instead of a thinking ape, we have merely one waiting for manna from above.

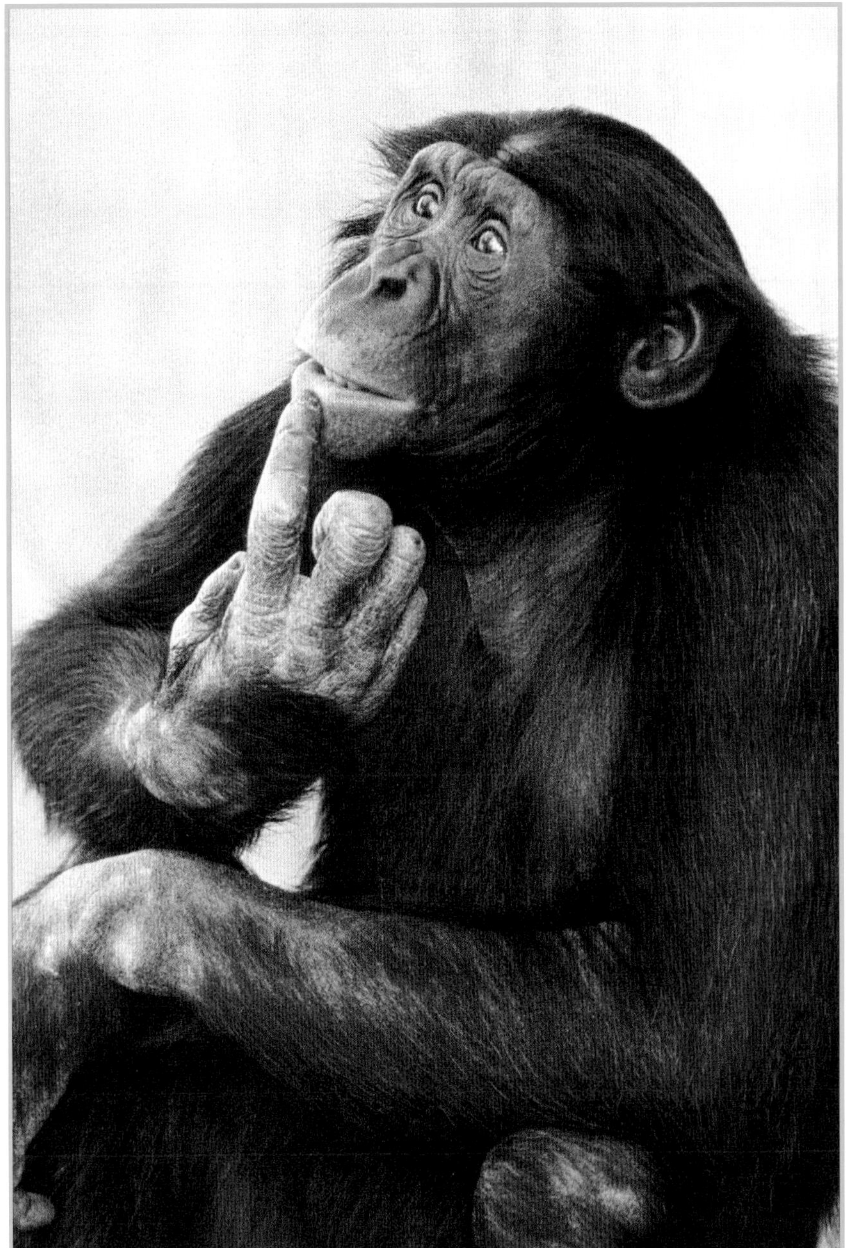

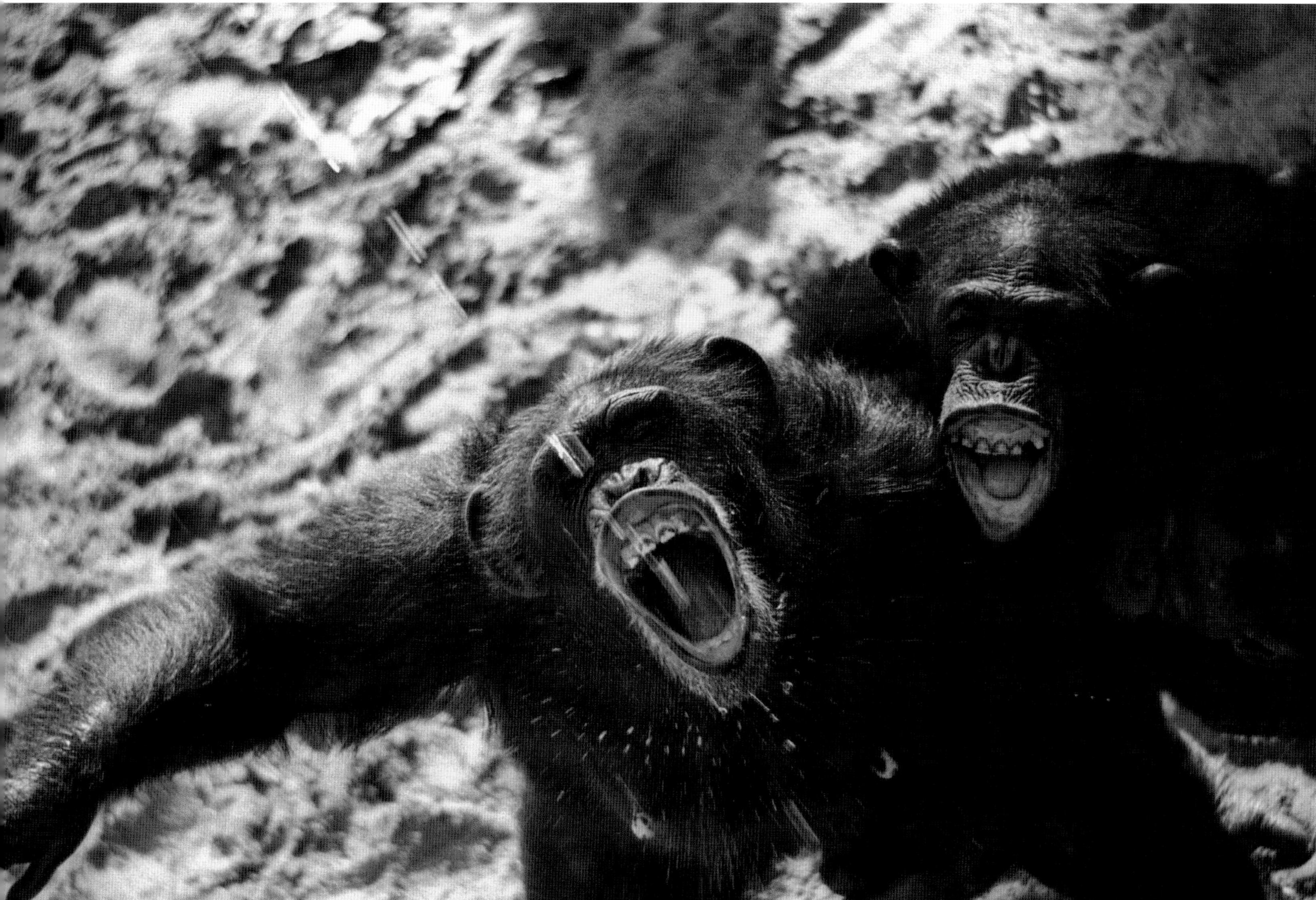

High tea

In the old days, zoos held so-called tea parties at which apes were seated around a table and trained to drink from cups. This was done to mock human cultural refinement and acquired taste. Having once suffered through afternoon tea in one of London's stuffiest restaurants, I can relate to the drill and am convinced that the Arnhem chimpanzees had more fun drinking tea from really high up. In fact, they were getting it from me, a habit I developed after having learned from an accidental spill out of the observation window how much they loved the taste. By the end of the day, we would hold up the teapot, the chimps would come rushing toward us, and we would empty the pot for the assembled tea aficionados, such as these two females. They were excellent tea catchers with their lower lips sticking out, although both the wind and the pushing and shoving often interfered, resulting in occasional tea showers.

Joint attention

When Nikkie fished berries out of the moat and started nibbling on them, others rushed over to see what he had found. Chimpanzees are naturally curious about what everyone else is doing, often collectively zooming in on the same item. Thus, if one of them turns to stare at something, others will do the same (see also pages 140–41). Once, I saw the entire chimpanzee colony silently follow the same unusual bird—a hawk flying by with a chicken in its talons!—after one of them had spotted the drama in the sky. I myself spotted it by following the apes' gazes. Gaze following is second nature to us, as it is for many primates. It is a critical step in human development, because paying attention to what others pay attention to enables children to get into the minds of others. Whether apes are capable of the latter is a topic of intense debate, but its precursor—joint attention—is easy to see.

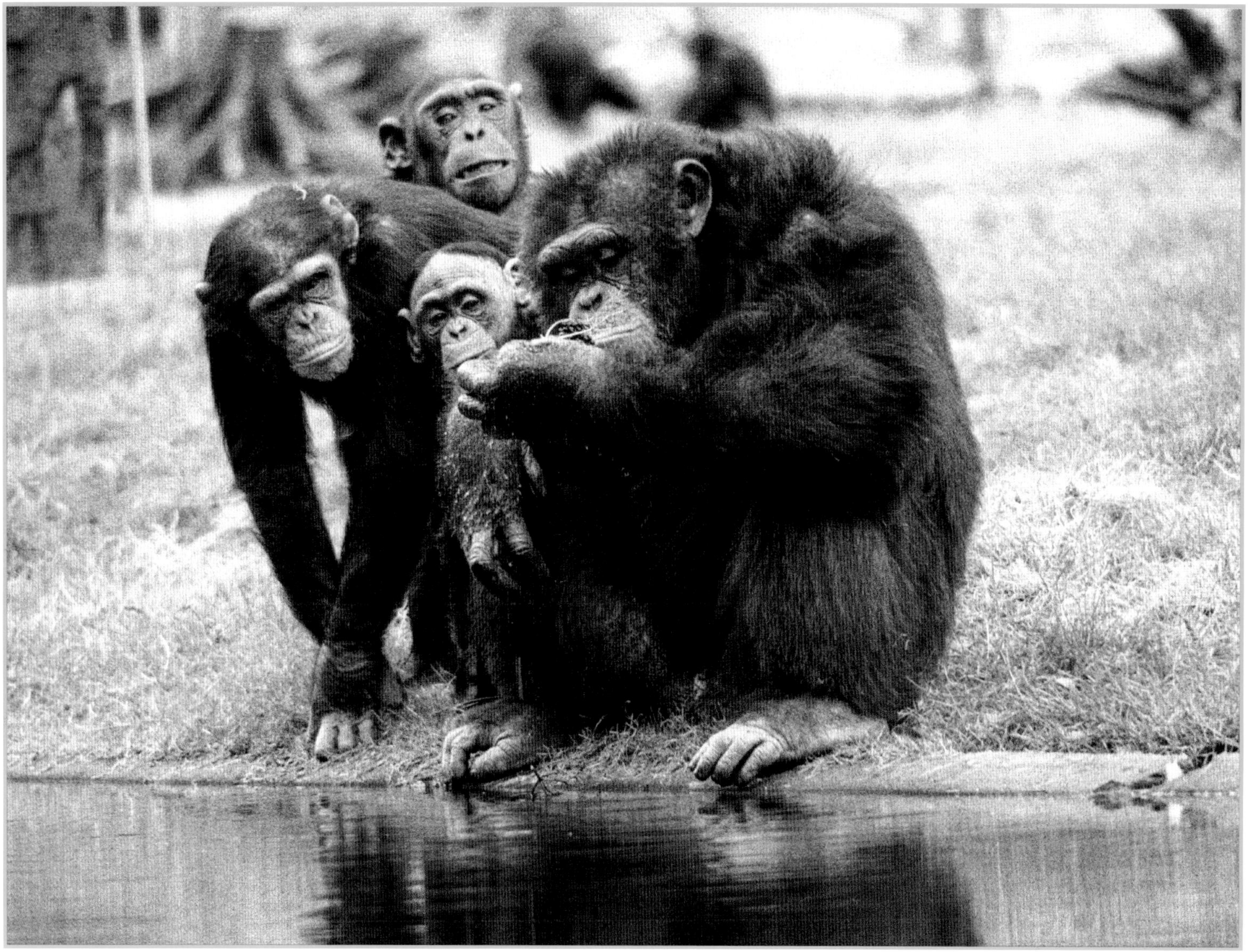

The ready foot

What looks like a young ape staring at a stick is actually a determined exploration of the environment. Moniek, who in this photo is only three years old, pushes on a dead branch with her left hand while holding her left foot ready in a grasping position. There are many things apes can do with their feet that we cannot: catching a falling object is one. When the branch snaps, Moniek will end up with half of it in her foot. For such a young ape this is quite a feat, as it requires her to predict the consequences of her own actions. Cause-effect understanding has long been attributed to apes, but it has to my knowledge never been captured in a single frame. Usually, one needs to see an entire sequence, such as the unsteady black-and-white movies produced by Wolfgang Köhler in the 1920s, in which chimpanzees try to reach a banana dangling above them. Provided with boxes and sticks, the anthropoids demonstrated planning and foresight. Köhler spoke of *insight*, a concept deeply upsetting to the psychologists of his day who studied learning: they saw animals as trial-and-error learners. Moniek shows how insight develops at an early age.

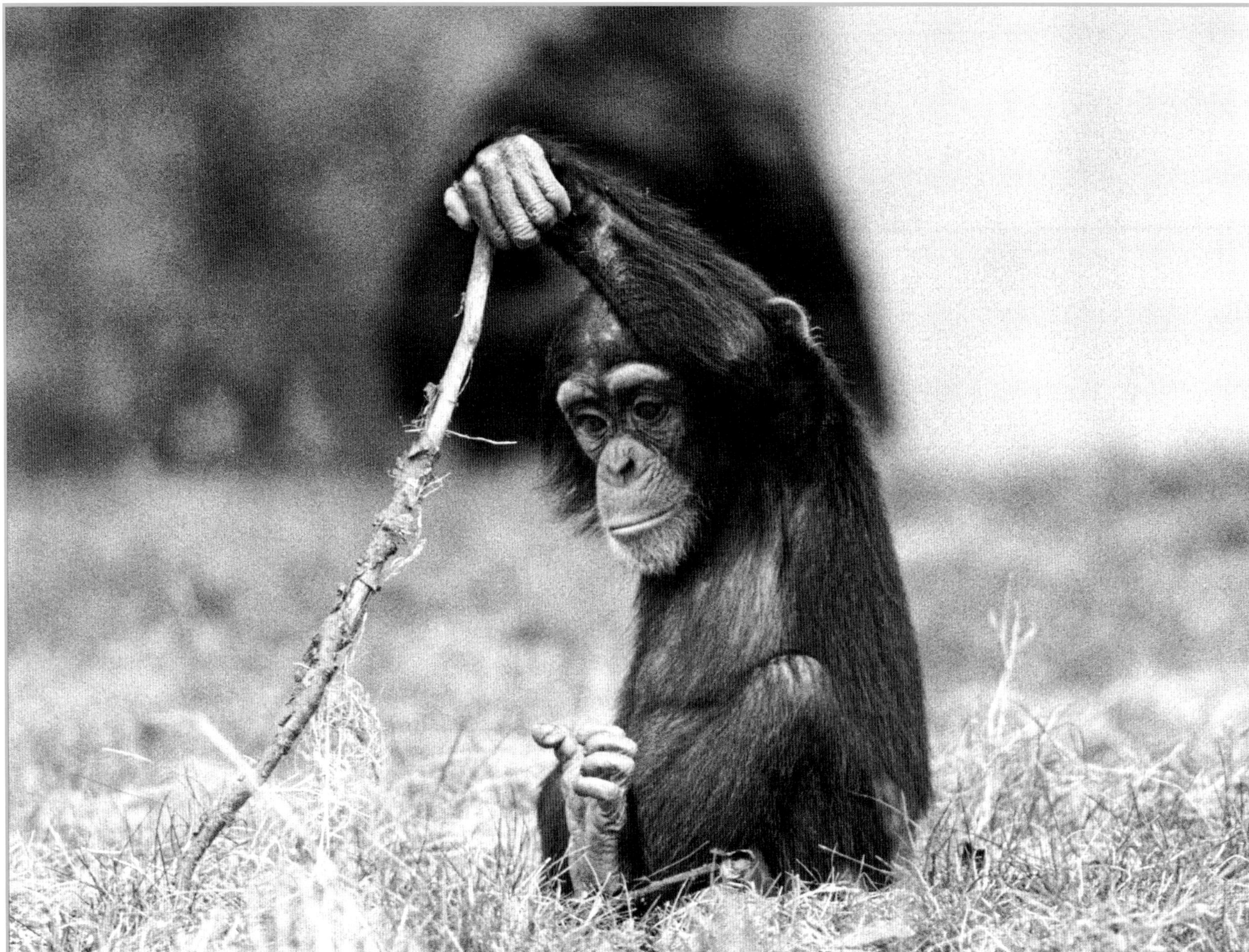

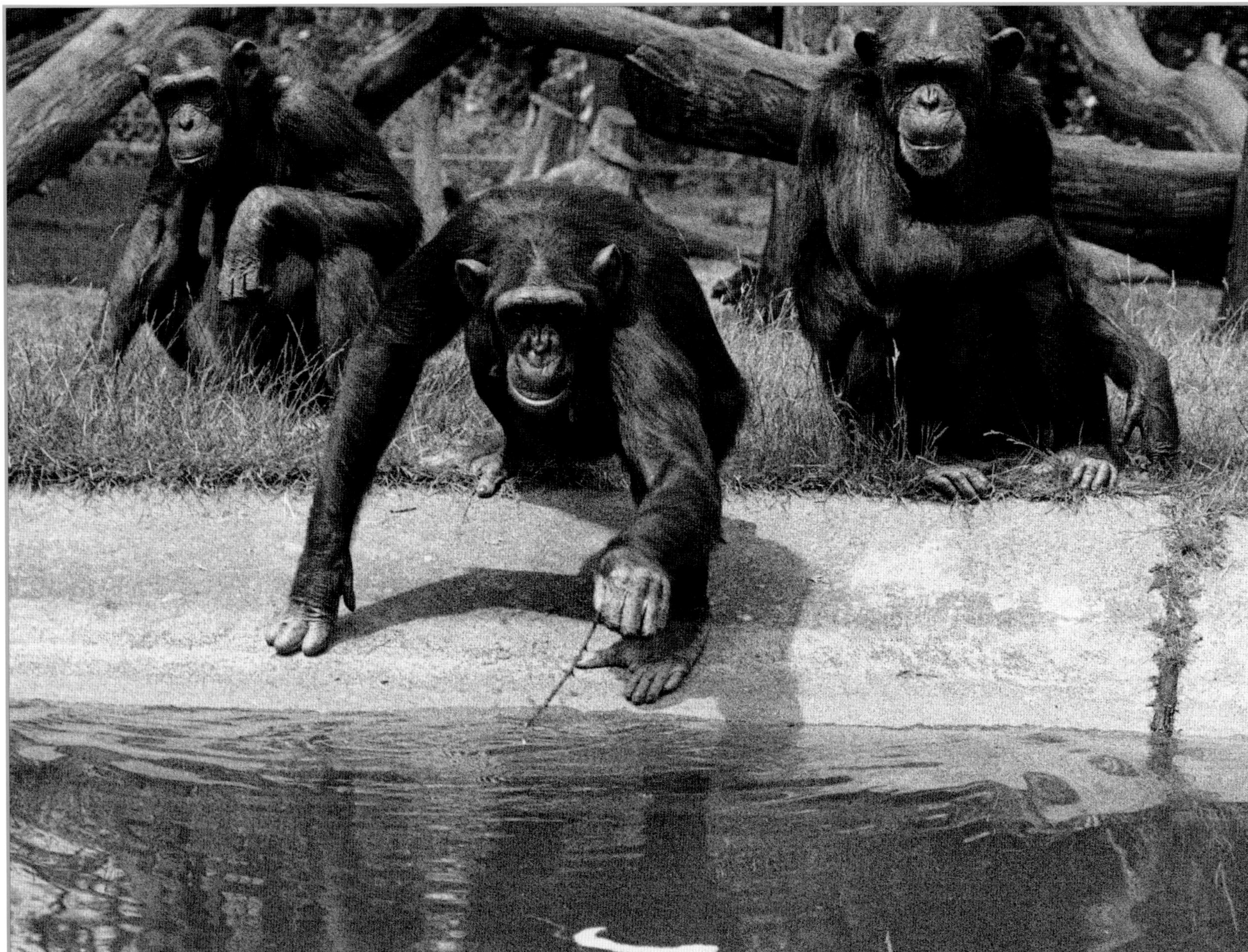

Tool use

The use of an object as an extension of oneself in order to reach a goal is limited to a few animals, including chimpanzees, capuchins, and orangutans. Both photographs show chimpanzees at the Arnhem Zoo manipulating a stick to fish something out of the moat. Both photos also illustrate the joint attention considered so important in primate social life (page 137). Bystanders have come over to follow the actions of the tool user. Rather than look only at the individual, they check on the target of the effort. This way, they gain important information about technical solutions of others, information that may help the cultural transmission of habits.

If a tool is inadequate—as in the large photo, where the twig is too short—the actor will break off its attempts and leave the scene to hunt around for something better. During this quest, the goal (the object floating in the water) needs to be kept in mind: a mental operation that has fascinated scientists over the ages, inspiring many experiments on tool use.

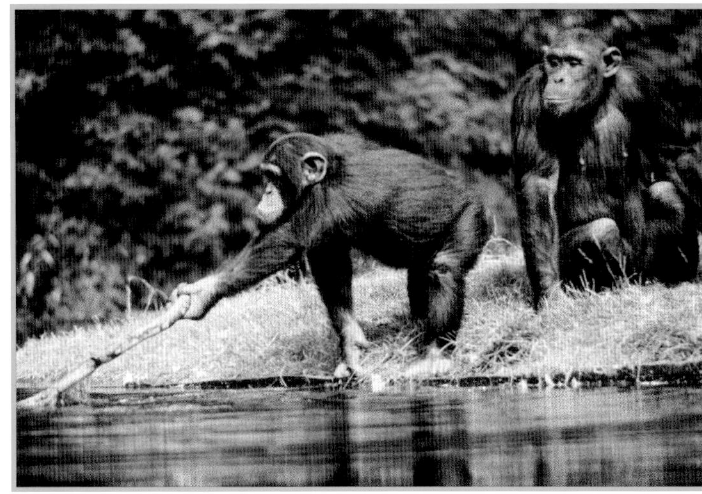

Roosje on the bottle

One of the most emotional moments of my career occurred in 1979, when we handed over a baby chimpanzee to an adoptive mother of her species. Following removal from the natural mother, who due to her deafness had failed to provide appropriate care (page 10), we had raised Roosje ourselves on the bottle (*Roosje* is Dutch for "Little Rose"). But since human-raised apes often turn out to be abnormal, we thought it would be better if a female of her own species took our place. For this, we chose Gorilla, a female fascinated by but unable to raise infants because of insufficient lactation. Every day, we demonstrated the feeding technique to Gorilla, praised her for holding the bottle without drinking from it, and let her get used to Roosje.

After many sessions with bars between Gorilla and us, the big day of actual introduction arrived. We placed the wriggling infant in the straw of a night cage and let Gorilla in with her. The small photo shows the adoptive mother intently staring into Roosje's face. Gorilla didn't dare touch the infant: it belonged to us. She approached the bars where the caretaker and I sat watching. First she kissed the caretaker, then me, glancing between Roosje and us as if asking permission. We both urged her, "Go, pick her up!" Eventually she did, and from that moment on Gorilla has been the most caring and protective mother one can imagine, raising Roosje as we had hoped.

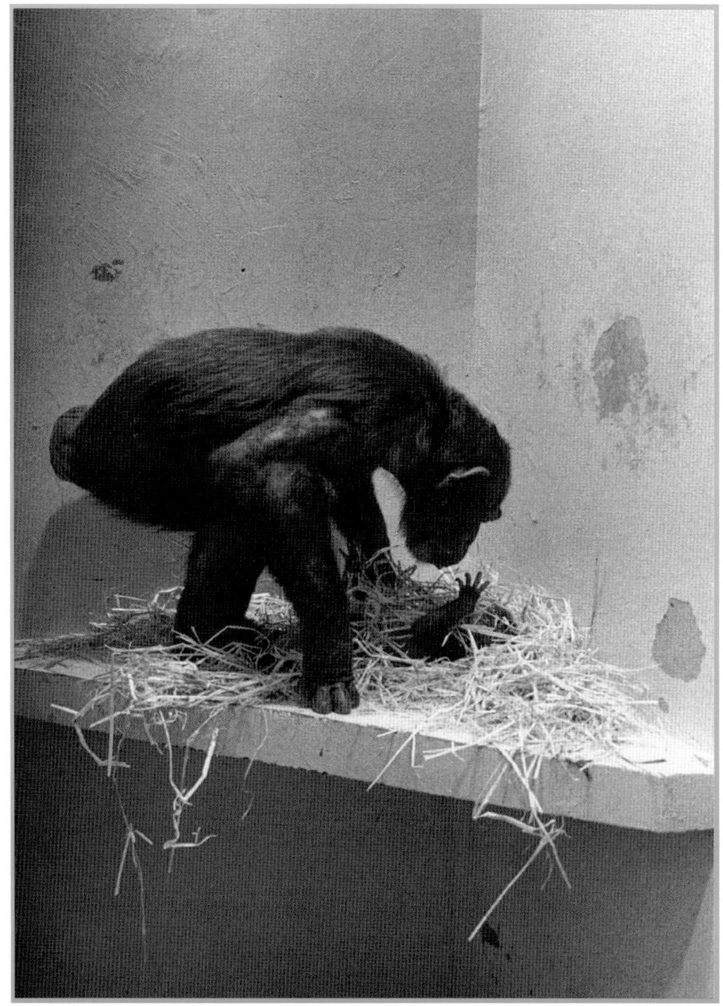

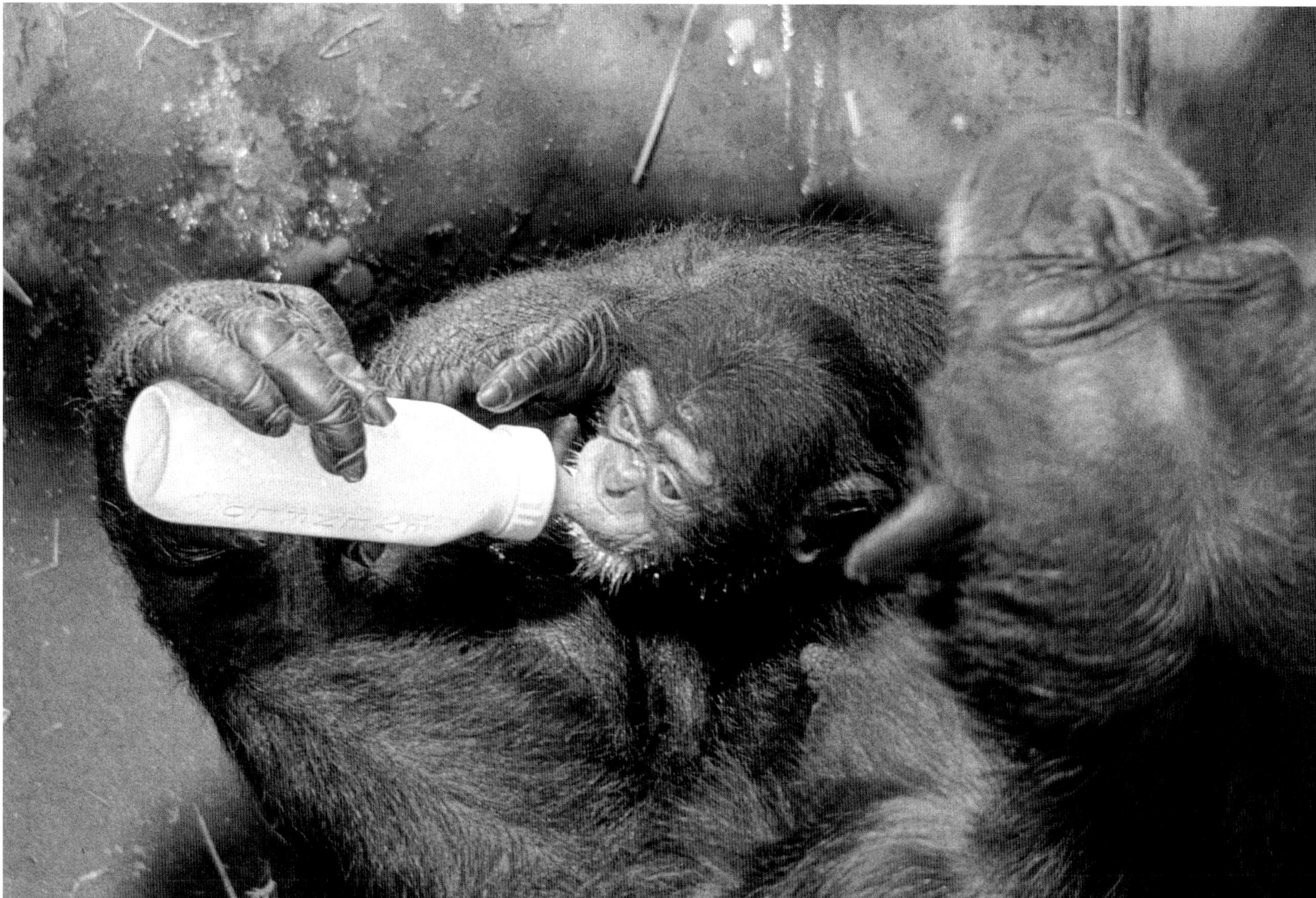

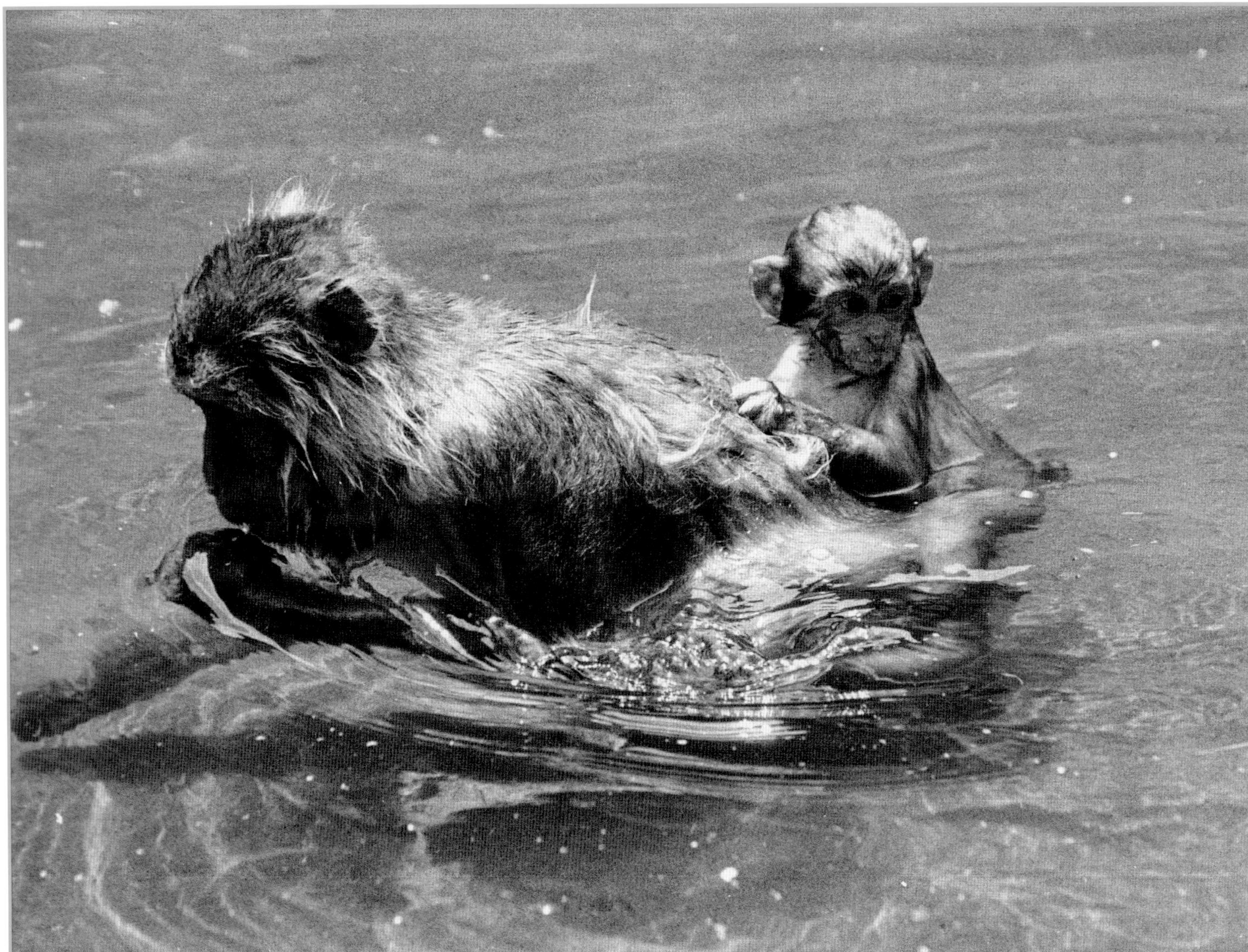

Holding on to mom

While the mother collects soybeans scattered in the water of a hot spring in the Japanese Alps, the infant clings to her back with its head above the surface. This position is not as logical as it may seem: park wardens often chase first-time mothers out of the water to keep them from accidentally drowning their offspring. Wardens don't need to take such action with older mothers, because the macaques gradually learn to prevent harm to their offspring in this unusual situation. It is unclear whether they do so by taking the perspective of their offspring (such as understanding that they need air) or whether they simply become more responsive to signals of distress. Either way, being in tune with the needs of the infant comes with experience. This infant is still at an age when panic will result in more clinging to the mom: the wrong thing to do under water. When young monkeys grow up, they learn to swim and no longer face these kinds of problems. Juveniles have great fun jumping in and out of the water like human kids around a swimming pool (pages 162–63).

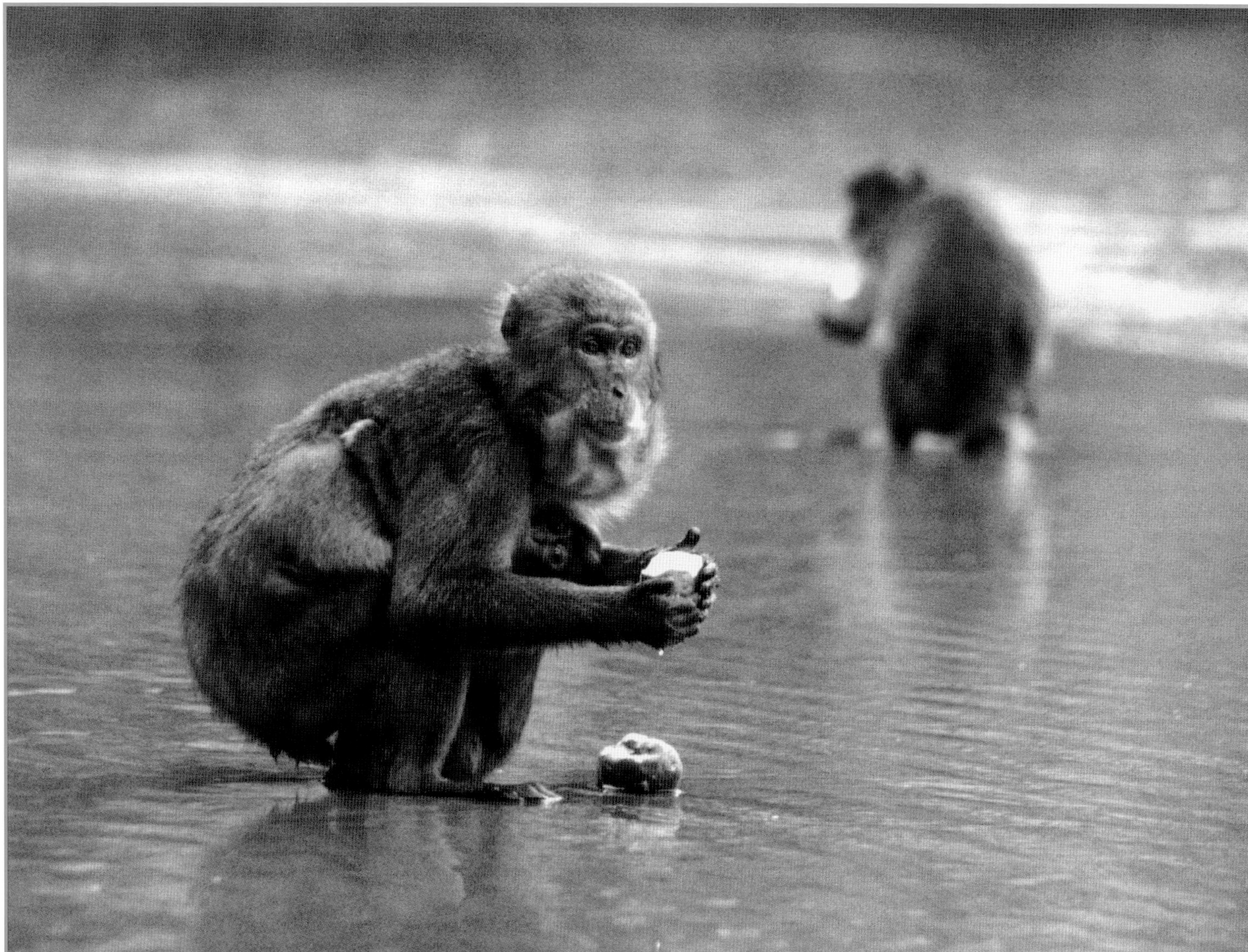

Cultured monkeys

In 1952, Kinji Imanishi predicted that if the essence of "culture" is the transmission of learned behavior from one individual to others, culture will be found in many animals. How right he was! For research on my 2001 book *The Ape and the Sushi Master*, I visited Koshima Island in the far south of Japan, where Imanishi's students had long ago found the first evidence for cultural transmission. On this tiny island, one juvenile Japanese macaque named Imo discovered that a good way of cleaning sweet potatoes was to bring them to a freshwater stream and rub them in the water. Soon Imo's mother and her peers followed suit, and within about five years almost all monkeys on the island washed their potatoes. The only category that seemed resistant to this cultural revolution, and never learned, was the older males. Years later, again led by Imo, the monkeys turned to washing spuds in the salt water of the surrounding ocean. When I visited in 1998 they were still washing them, even though Imo and her cohorts had died off.

Nowadays, it is impossible to obtain dirty potatoes in Japan, but the monkeys keep washing them. Consequently, the Japanese investigators have proposed a new label: "seasoning behavior."

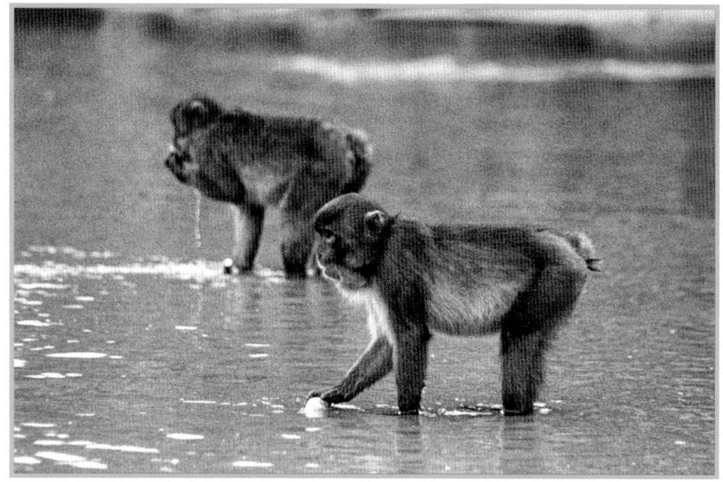

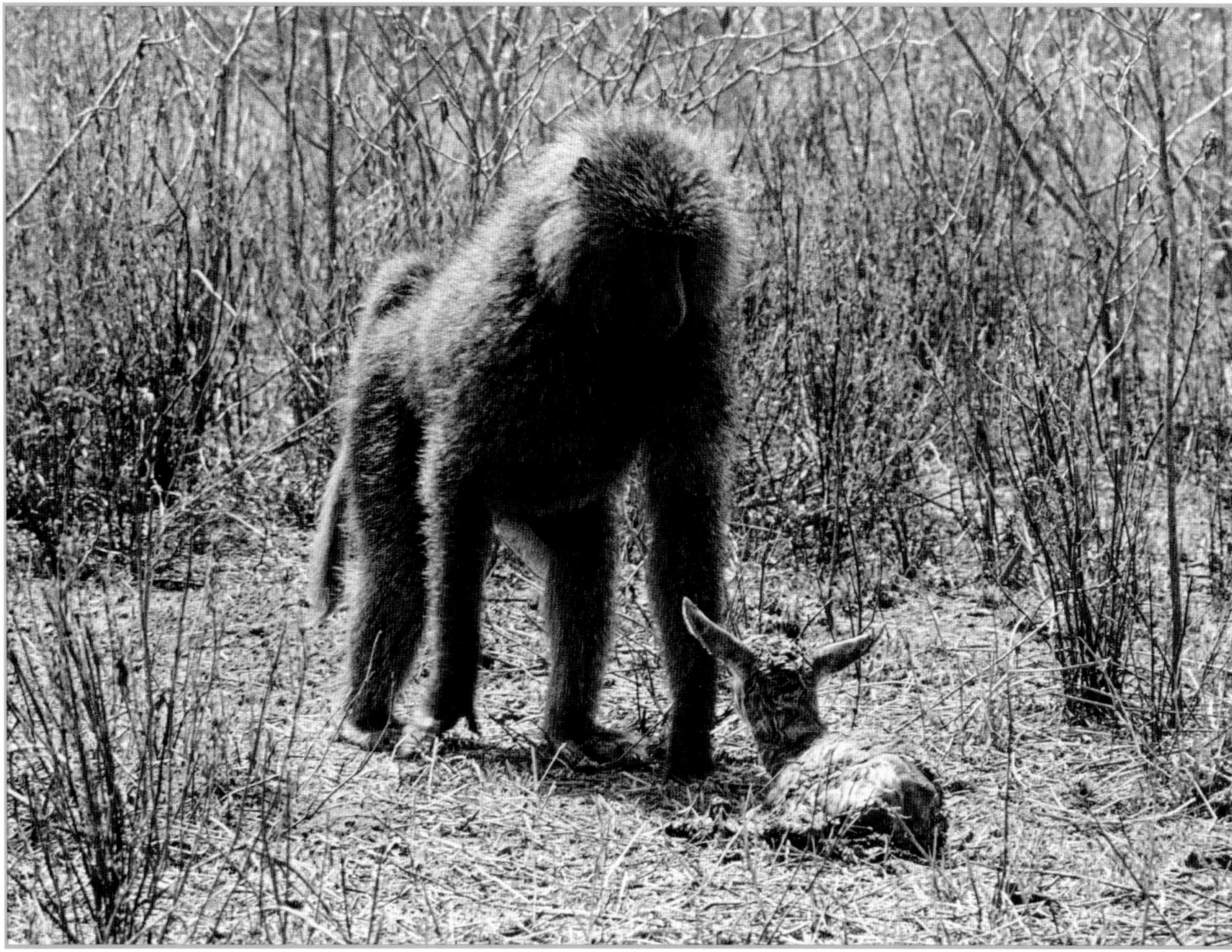

Strange encounter on the savanna

Baboons are known to kill and eat the calves of Thomson's gazelles. I therefore expected a bloodbath when I witnessed this rare encounter on the plains of Kenya. We were traveling with a troop of olive baboons, which were very scattered at the moment an adult female discovered a still-wet newborn Tommy calf in the grass. She faced the calf for a few seconds, then sat down and looked around. Female baboons never kill the calves: it is only the adult males, with their enormous canine teeth, who do so. The female kept staring in the direction of a couple of males in the distance but didn't call out or draw their attention. After a few minutes, she moved on, leaving the lucky calf alone.

Lack of effective communication about a food source is unsurprising for a species, such as the baboon, in which sharing of food is absent. In this they differ greatly from, say, the chimpanzee. Chimpanzees call loudly when excited, and a female chimpanzee in the same situation might well have attracted others. The caller would likely have received a share of the meat, which is far from guaranteed in the case of a female baboon.

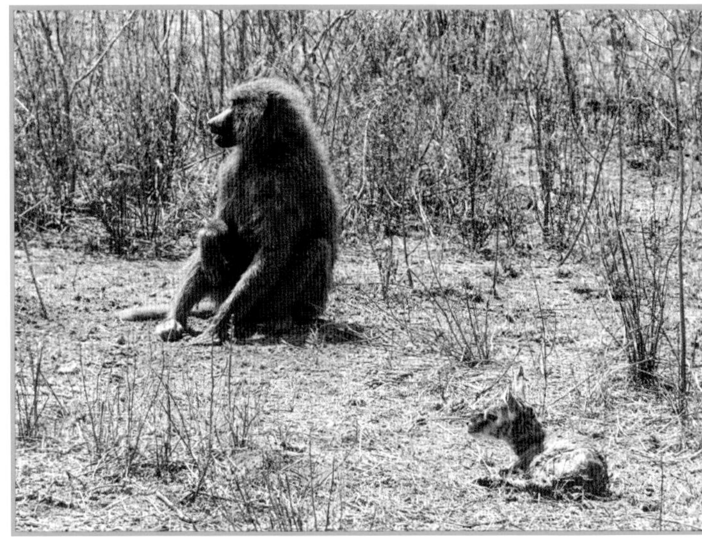

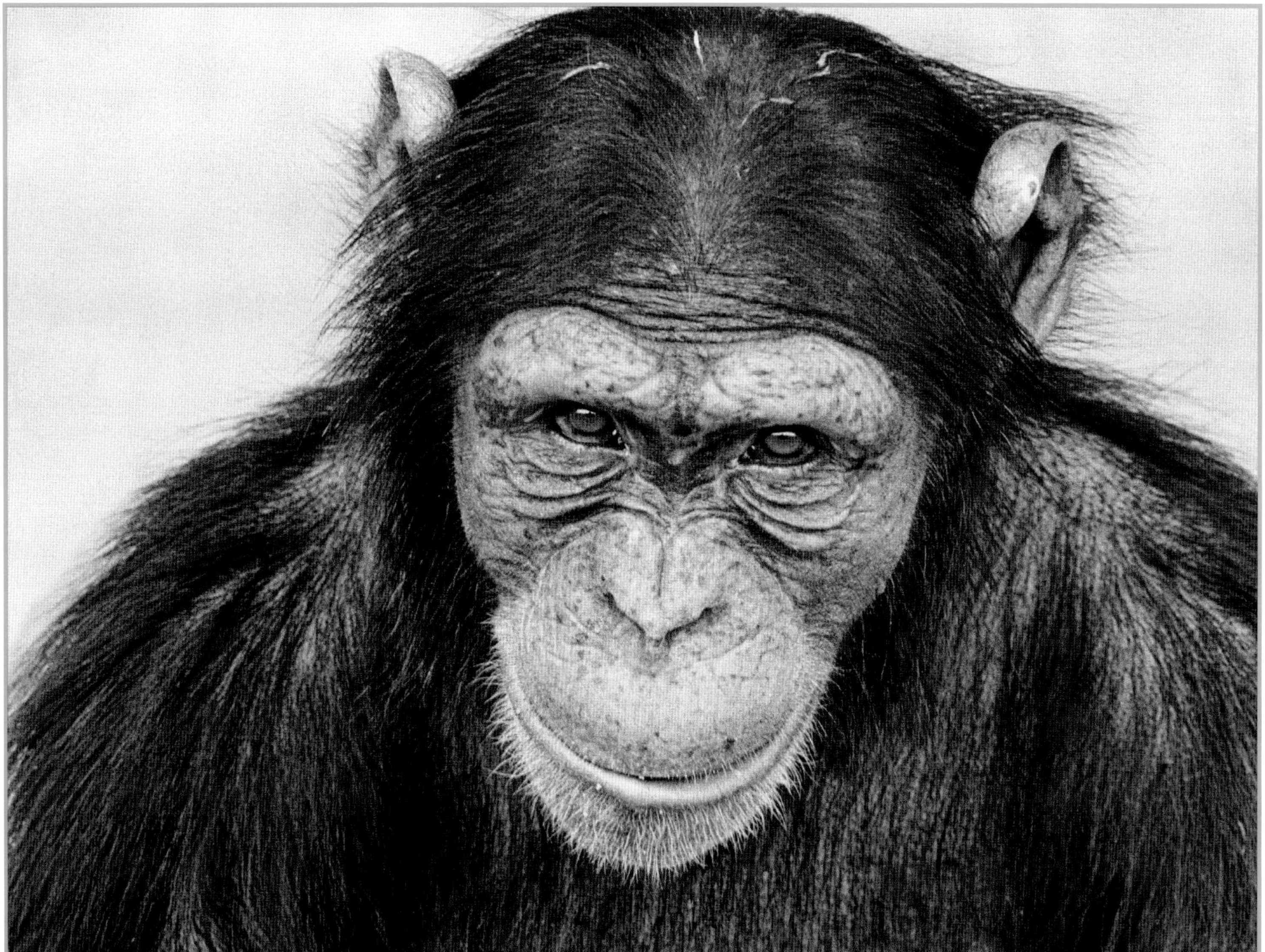

The naughtiest chimp

Georgia hangs several meters above the ground, right underneath me, staring with her usual intensity straight into my camera in search of herself. Before we built a tower, I used to watch the chimpanzees at the Yerkes Primate Center Field Station leaning precariously over the wall of their enclosure. Georgia would climb up to me in order to play with me and check out my lens, which serves as mirror to her. It has long been known that chimpanzees recognize themselves in a mirror. This explains why they are more intrigued by their reflection than, say, monkeys or dogs, who see only a stranger.

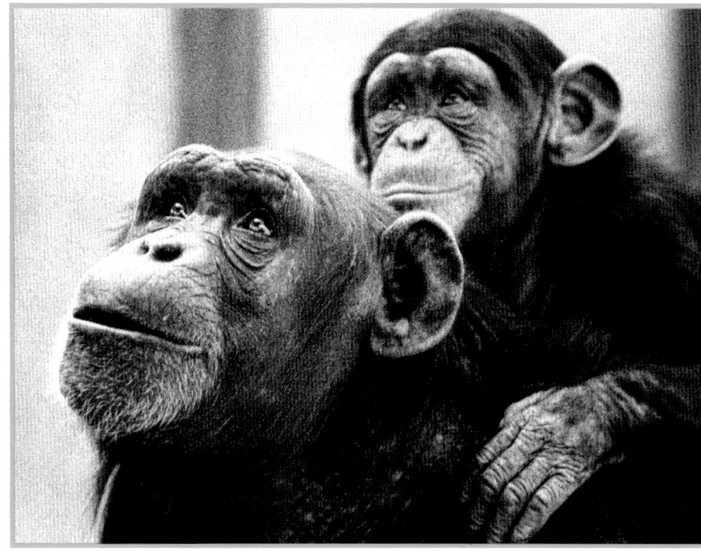

Georgia is now a mother (small photo) and has mellowed, but I have known her since she was little and consider her to be the naughtiest chimpanzee I have ever met: she is a certified troublemaker and social rule breaker. Only six years old in the large photo, she had already developed the habit of surprising visitors by spitting a mouthful of water. She would suck water from a faucet as soon as she saw strangers arriving in the distance, then mingle casually with the rest of the colony until she had the people close enough for a surprise shower.

Gentle baby-sitters

Kalind (right) is putting on an asymmetrical face while on the lookout with his play partner, Lenore. Seven years old, Kalind is ready to enter puberty, whereas Lenore is only a two-year-old infant still being nursed by her mother. Not just among primates, but in many other animal species as well, grown-ups must be careful when playing with the young. Young apes love to play. They often invite adults to join in, jumping on top of them (pages 160–61), pulling at them, wrestling with a partner many times their size. The younger partner is permitted to go all out, hitting and gnawing at full force, whereas the older partner is expected to adjust, hold back his or her strength, and be gentle. Play thus provides important learning opportunities for older juveniles: they learn how they can harm others and how to avoid it. If hurt, the younger play partner will let out a loud scream, which will put an end to all the fun. There may be grave repercussions for the older partner if the mother steps in. Kalind had learned all of these lessons, however, and was extremely careful with Lenore, who couldn't get enough romping around with her favorite uncle.

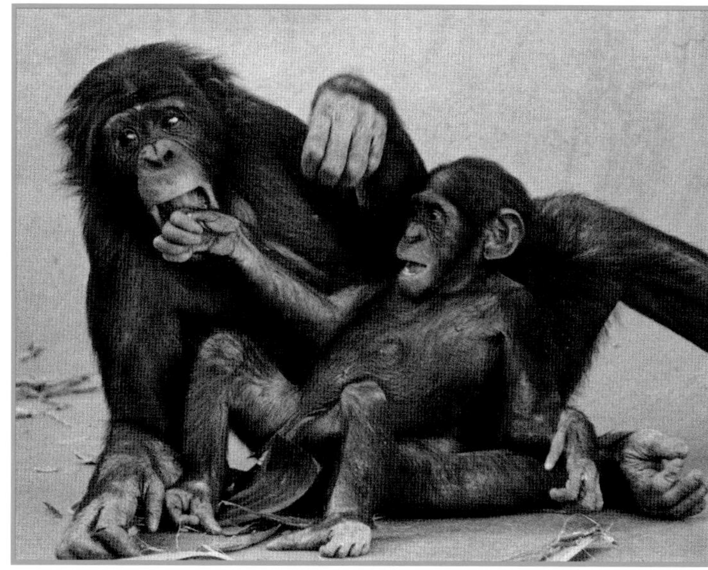

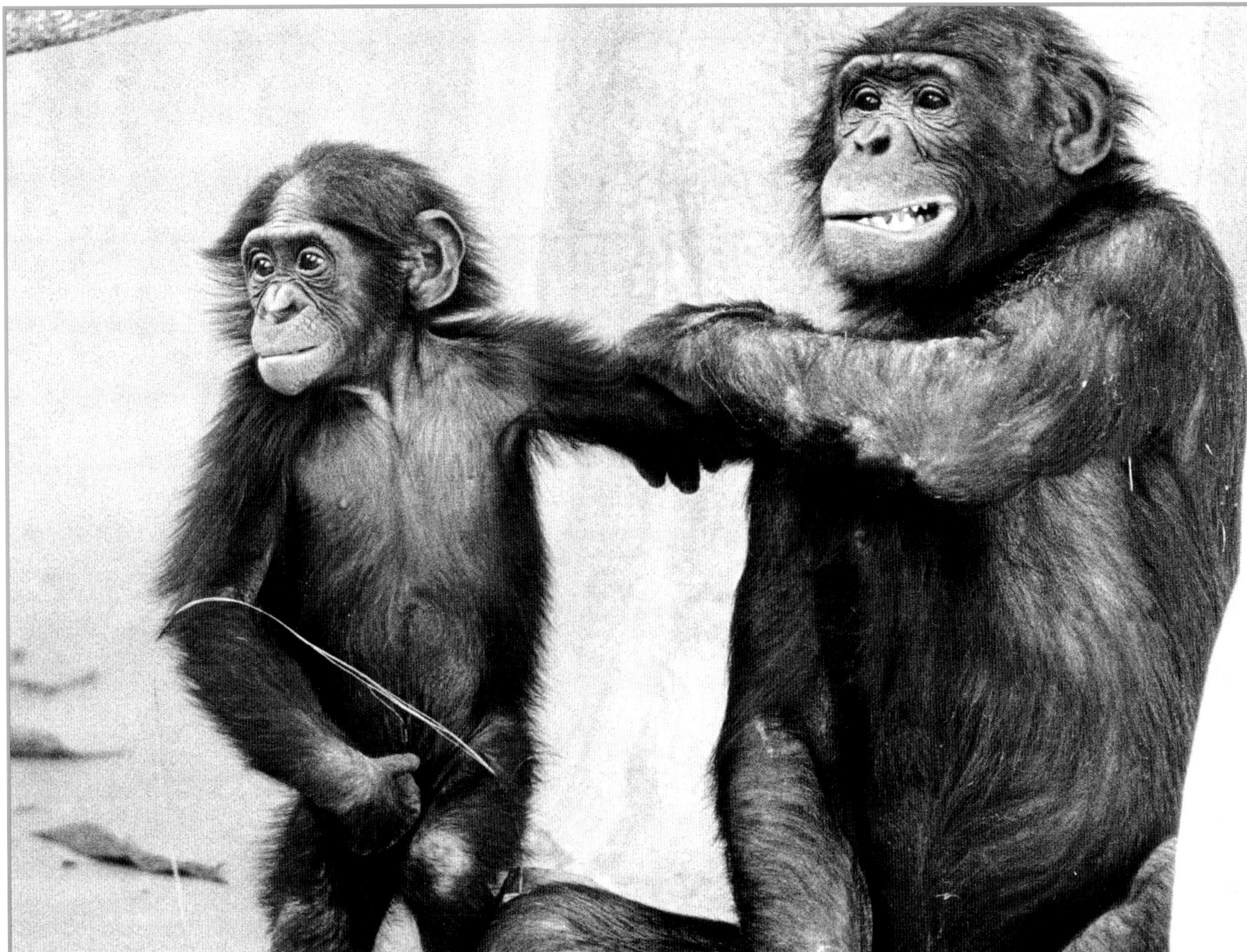

Upside-down play

A hallmark of human evolution is that we are neotenous: we preserve juvenile, even fetal, characteristics into adulthood (such as naked skin and relatively large heads). The same has been argued for bonobos in comparison with chimpanzees. For example, the little white tail tufts of young chimpanzees, which they shed at around age four, are retained by bonobos into adulthood. Another juvenile characteristic is playfulness, which both humans and bonobos possess all their lives. Human art, dance, and music, not to mention sports, attest to our continuous playfulness. Bonobos may not have the same outlets, but they have plenty of games and keep playing these even when fully grown. Here an adolescent male (left) plays with an adult female. Incredibly—for any adult primate—the female is so completely engaged in the game that she hangs upside down while grasping for the male's feet.

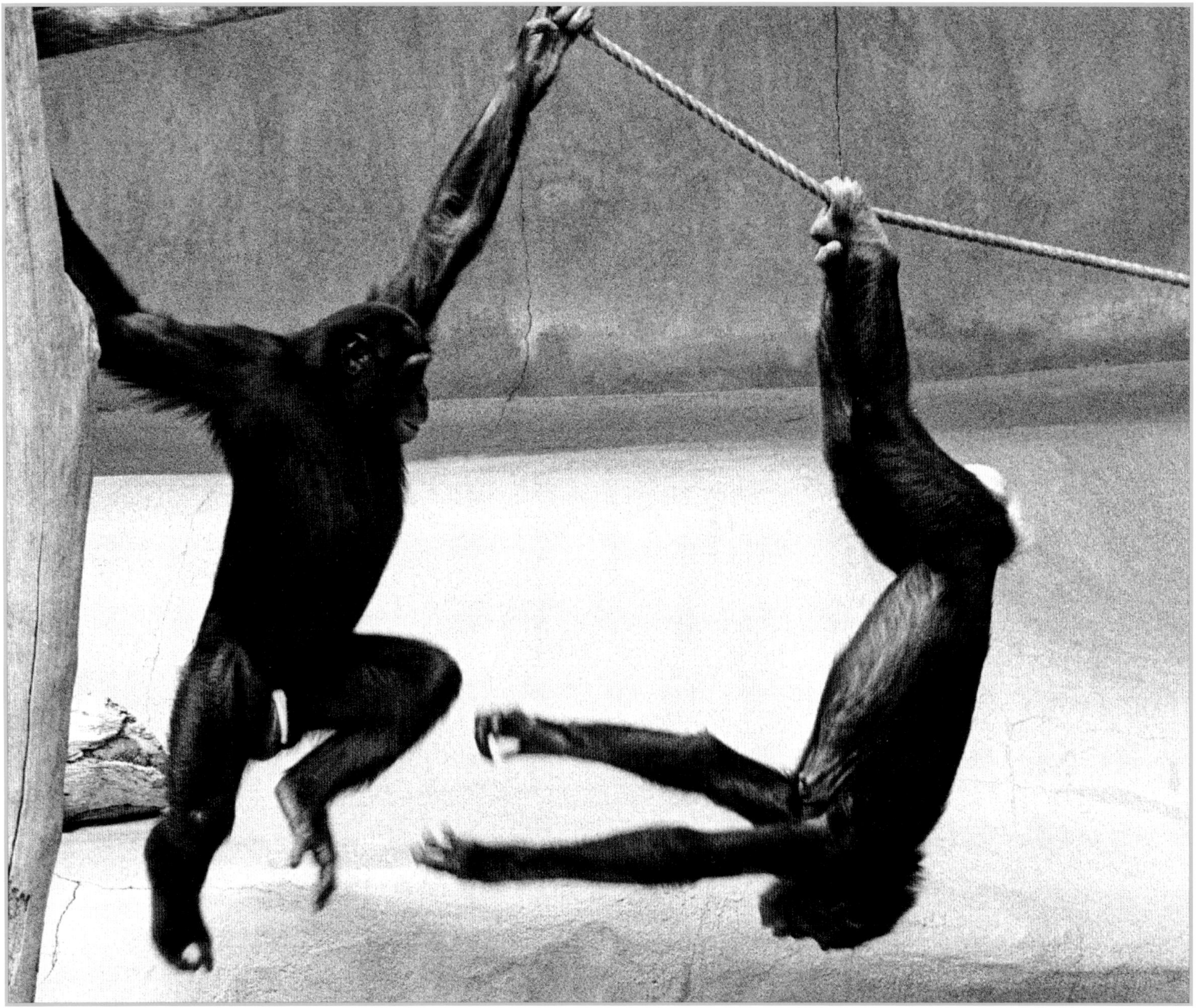

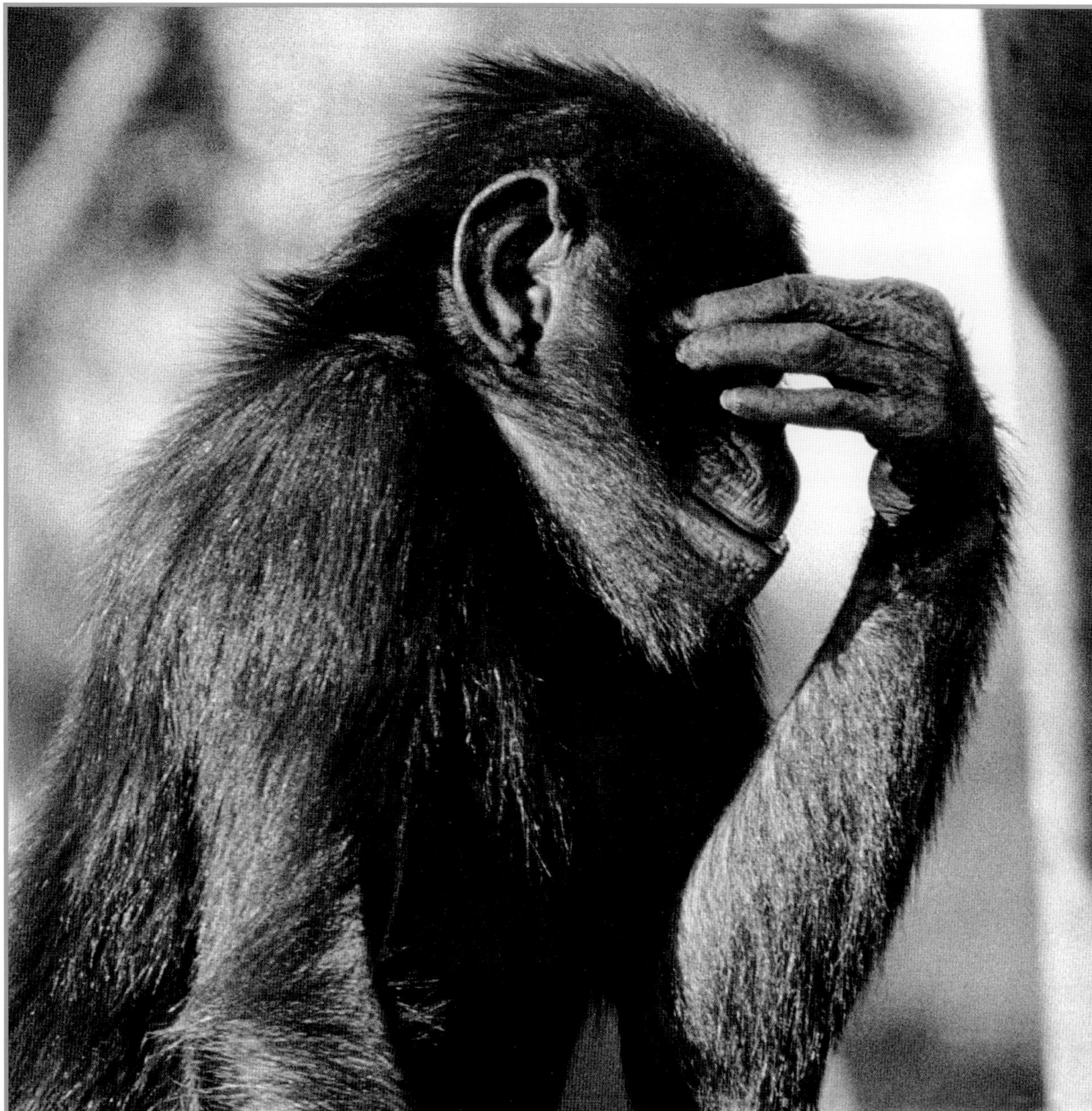

Blind man's buff

Juvenile bonobos come up with new games all the time. One game they love is blind man's buff. Two females of five and six years of age played the game at its most sophisticated level, climbing so high up in the familiar wooden frame of their enclosure that a fall could be dangerous. They would stick two thumbs in their eyes, sometimes a thumb and index finger of one hand. Or they would cover their faces with a banana leaf. They seemed to keep their eyes firmly covered while carefully, step by step, stumbling over the logs, grasping around blindly, or feeling around with a foot, trying to remember, it seemed, which path led where. This was a truly amazing feat of self-handicapping: a game calling upon a mentally stored map of their three-dimensional space. The self-imposed rule of the game—"I am not allowed to watch"—was less strictly followed by two males aged three and four. These rookies would poke only one finger in one eye or, as in the right-hand-side photo, place a whole arm over their eyes, moving the arm up and down depending on how secure they felt.

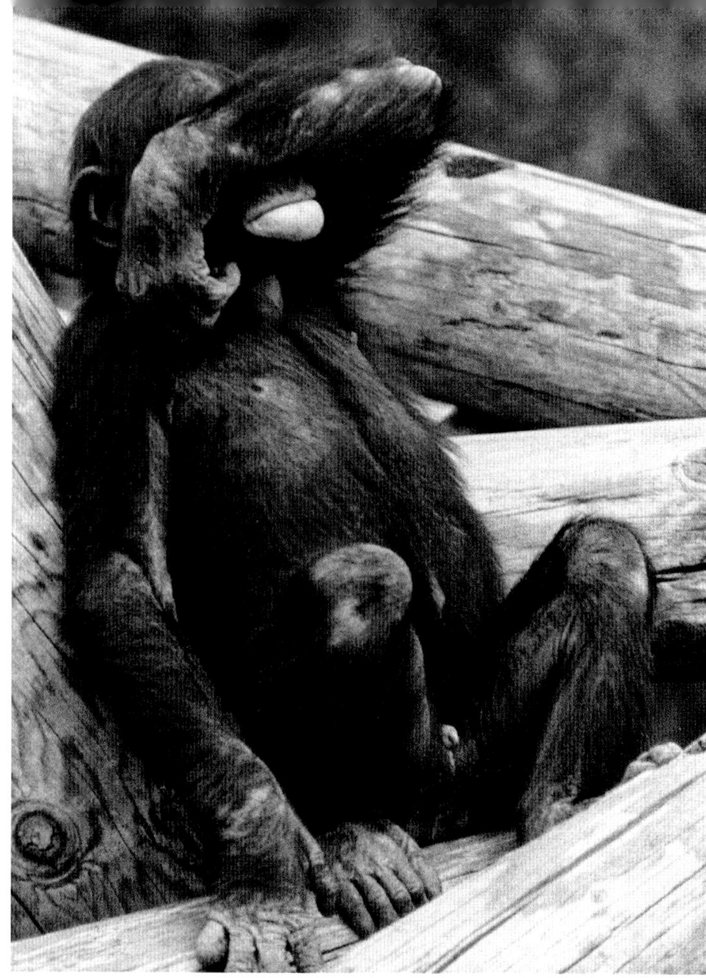

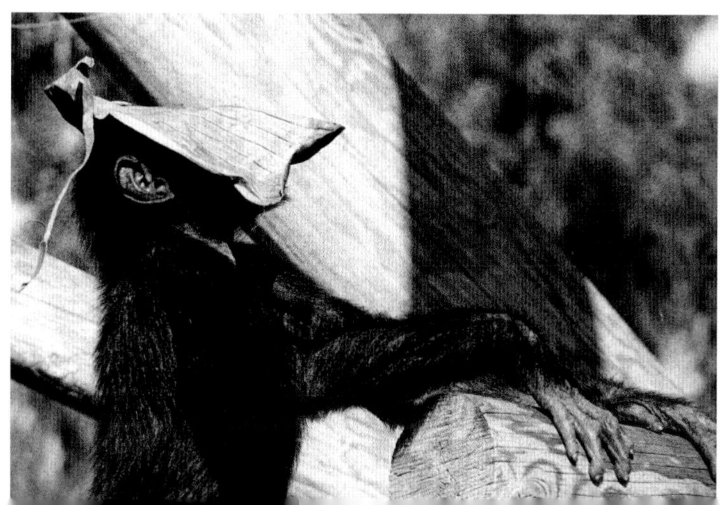

Getting stoned

The bonobos at the San Diego Zoo have discovered a natural drug, which makes them see a spinning world. This young male spins around and around while holding a rope, twisting at an ever-increasing speed until he can no longer continue. Then he releases the rope and sits passively staring in the distance, his mouth open. After a while he starts over and goes on until he is dizzy, again sitting with this dopey expression on his face. He must enjoy the experience; otherwise why would he do this repeatedly? The activity shows that bonobos like to play with their perception of the world, a tendency also underlying their game of blind man's buff (pages 156–57). Apparently, the natural curiosity and intelligence of the bonobo is not satisfied with the surrounding reality but seeks to enrich experience by transforming it. This is another reason, apart from the bonobo's free-loving sexuality, to call them the hippies of the primate world.

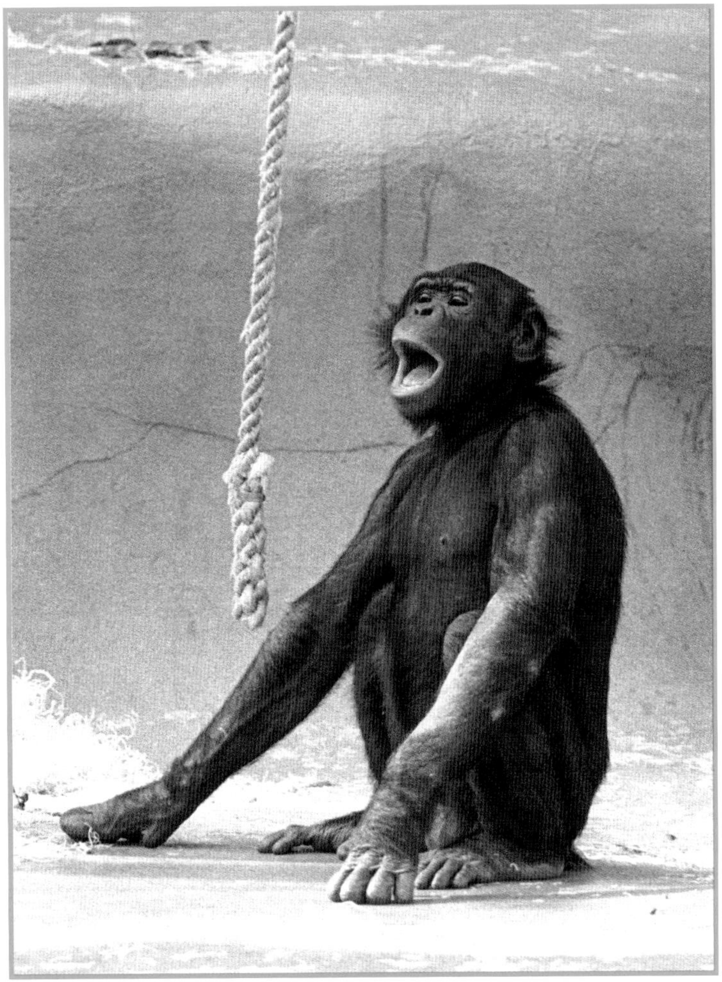

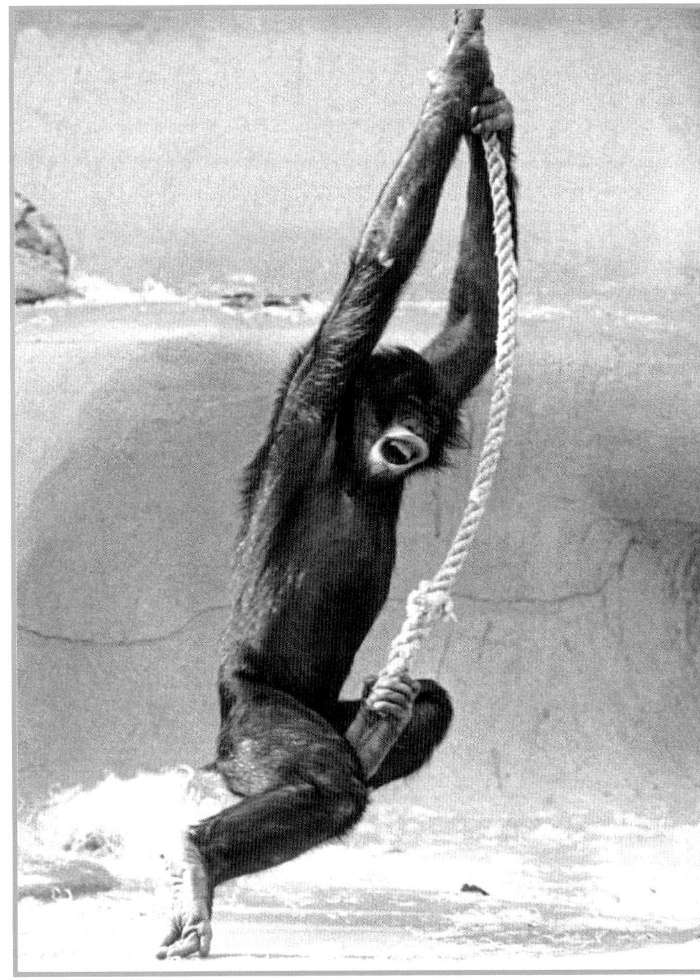

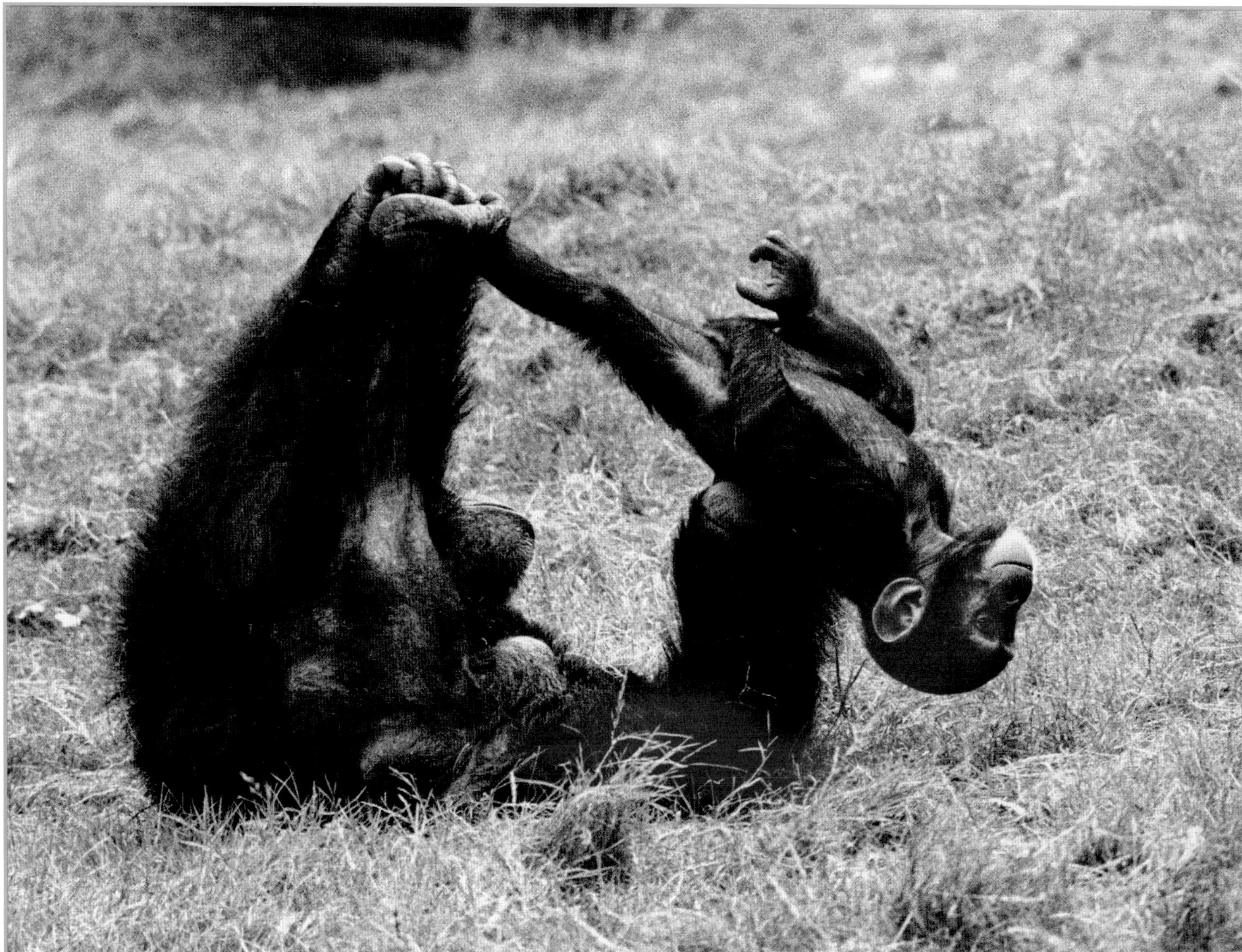

Narrenfreiheit

Liberties taken by the young in a primate society contrast with the strictly regulated life of adults. The youth are like court jesters who in times past could ignore the royal hierarchy and say and do (almost) whatever they wanted. In German this is known as *Narrenfreiheit* (*Narr* is a clown or jester, and *freiheit* is freedom). Thus, while an adult male and female chimpanzee sit on the grass early in the morning, a juvenile female decides to use the male's back as a trampoline (small photo). It seems like fun, but it is doubtful that the jumped-upon male enjoys it as much as the juvenile does. He won't do anything to stop it, though. The male in the large photo is actively engaged with Moniek, the same juvenile seen in some other Arnhem pictures (pages 34, 139). He gives the impression of tearing her apart as he lifts her up in a kind of airplane game that shows the complete trust that can develop between a vulnerable girl ape and the fearless leader of the group, Nikkie. When young, Moniek showed great fascination with big strong males, actively seeking out games with them.

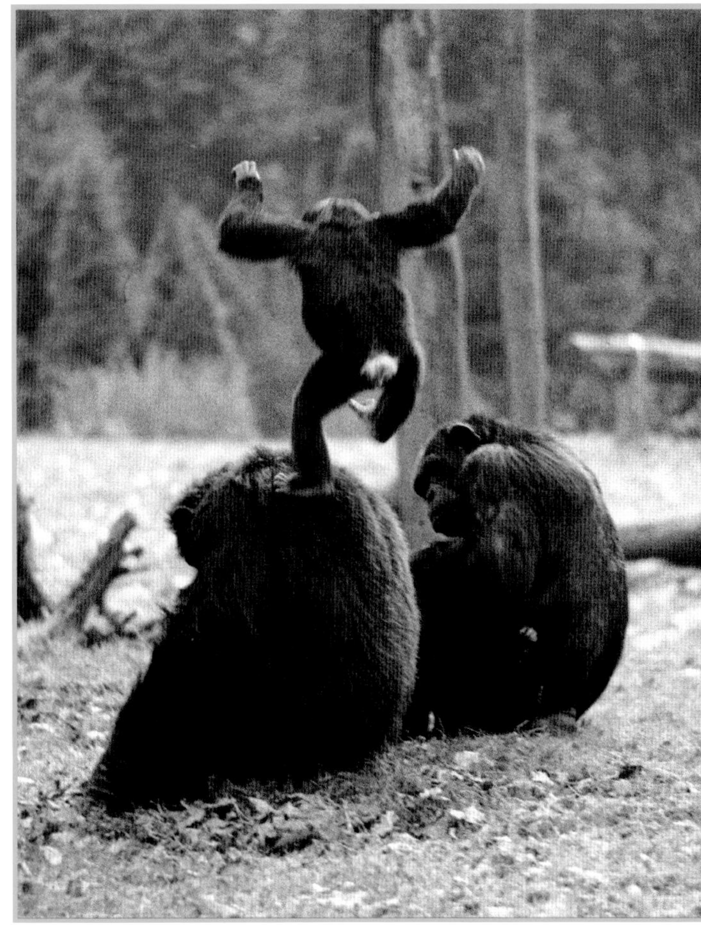

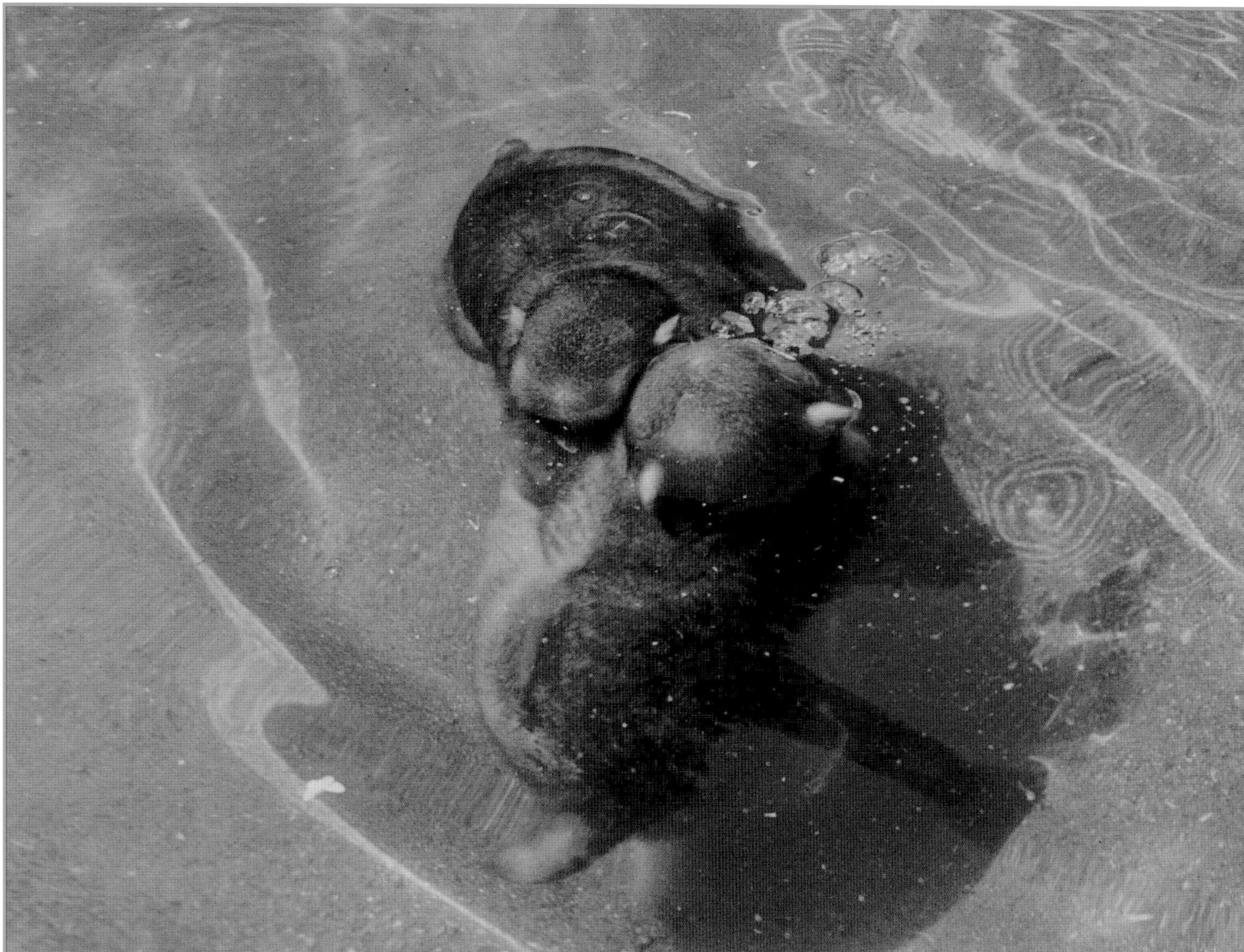

Underwater wrestling

In contrast to apes and untrained people, monkeys are excellent swimmers, as is shown by these two Japanese macaques in Jigokudani Park. Famous pictures have been taken here of monkeys with snow-covered heads warming themselves in the park's hot springs. But the monkeys use the pools during the summer, too, mainly for diving and swimming. These juvenile males kept ducking under water, wrestling and pulling at each other, then bursting out of the water for air (small photo). It made me wonder whether each monkey knew that the other needed as much air as it did. Perhaps preventing the other from getting that breath was part of their game. Occasionally, I saw a fight erupt after playmates had surfaced, maybe because one had held the other under for too long. But most of the time it seemed like great fun, with monkeys submerging and emerging together in a playful water ballet. That they can feel at home in a milieu totally unlike the one their species adapted to attests to the adaptability of these primates—an adaptability that humans have taken to an extreme.

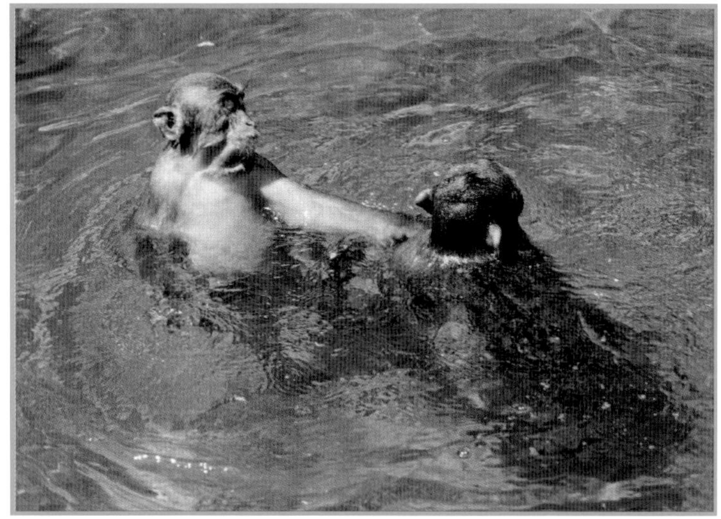

Sweet dreams

Of all nonhuman primates, Japanese macaques live in about the coldest climate, and mother nature has covered them with an unusually thick coat of hair. Japan is not wintry year-round, however. In the heat of the summer, the monkeys must be uncomfortable with all that fur. An adult male has found a cool rock in the shade to rest his head on. We don't know what male monkey dreams are made of, but I bet they include food and females. This male was so soundly asleep that he didn't wake up even when I approached for a close-up. But perhaps he fooled us and didn't care about cameras. When, a few minutes later, another male monkey made a branch-shaking display in a distant tree, this one jumped up to check out the scene. Keeping an eye on rivals is a major occupation of the male primate, far more important than observing the behavior of visiting photographers.

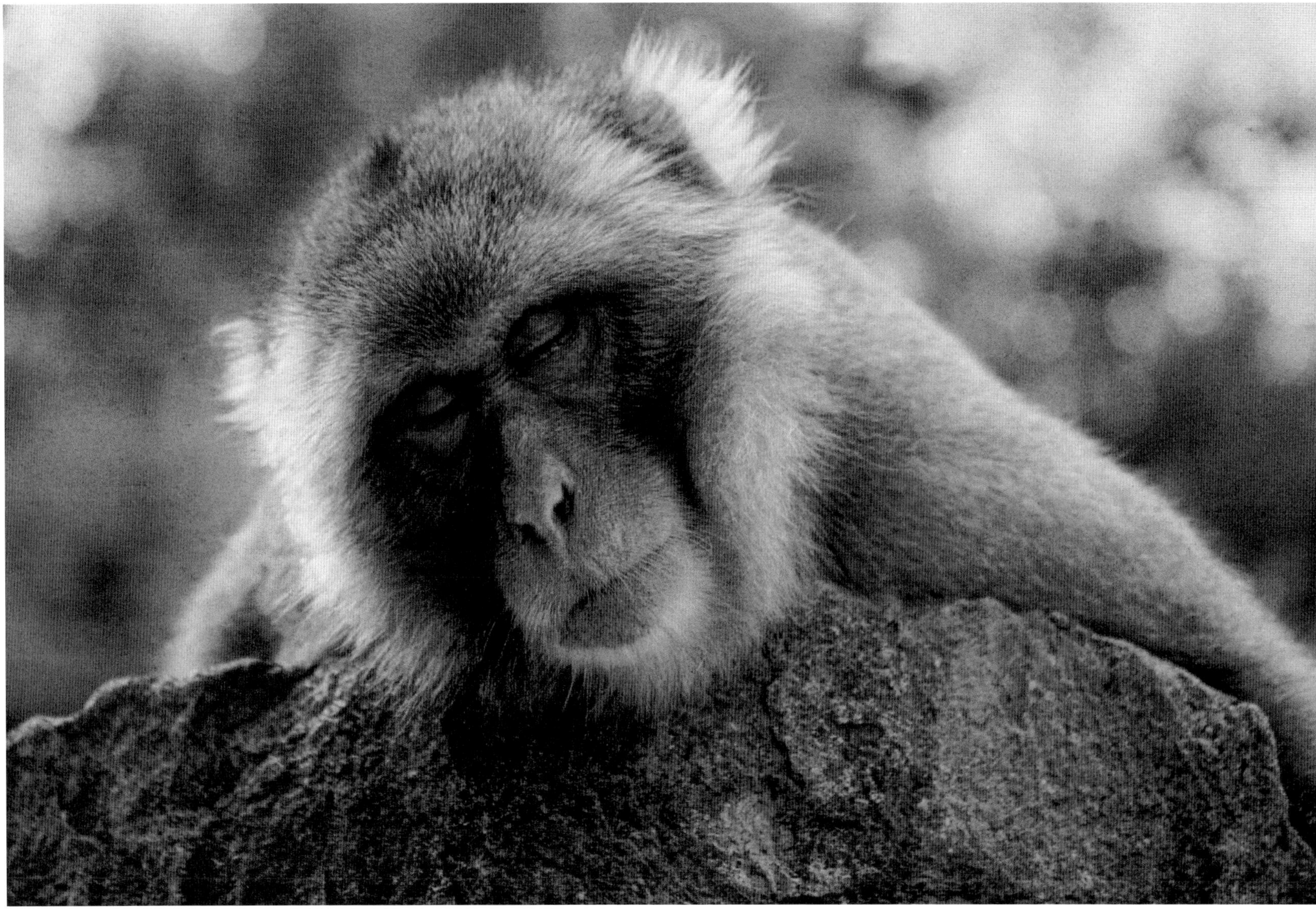

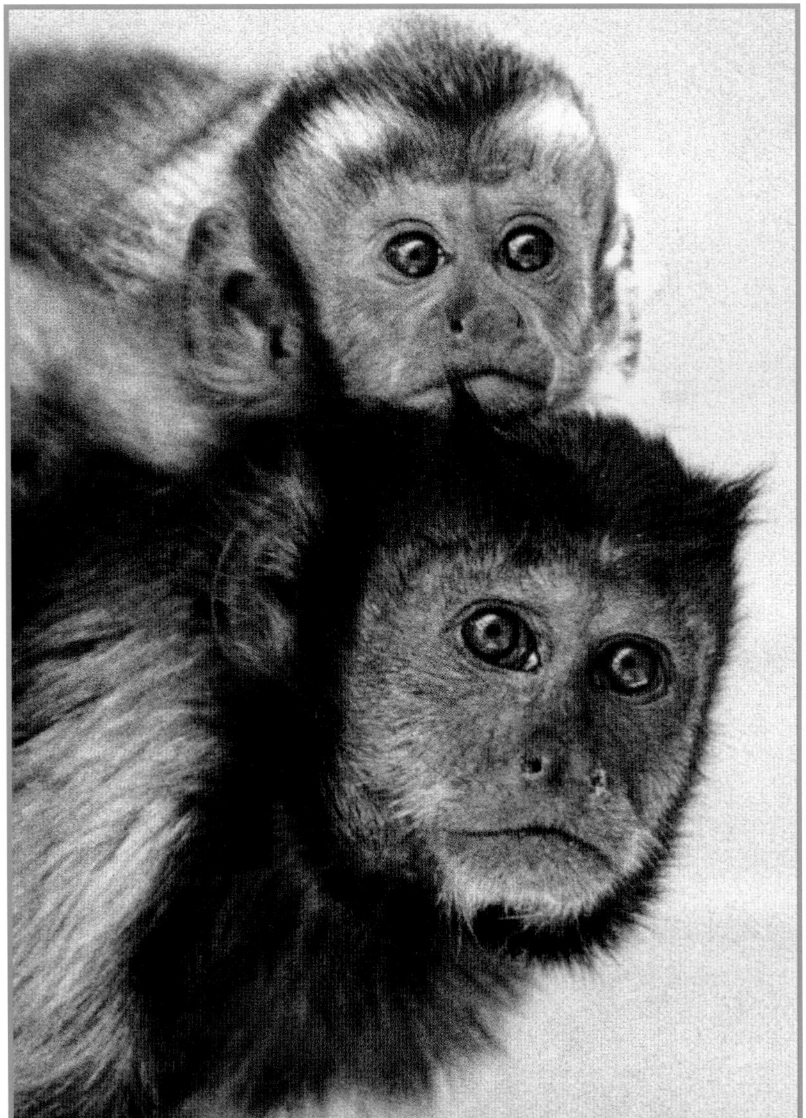

Photographic Locations and Species

Locations of Photographs

Burgers Zoo, Arnhem, the Netherlands

Pages vii, 10, 34–35, 36–37, 54–55, 74–75, 84, 87, 97, 98, 105, 106, 108–9, 118, 134, 137, 139, 140–41, 142–43, 160–61

Farm near Madison, Wisconsin, United States

Pages 15, 95, 111, 120

Fazenda Montes Claros, Brazil

Page 116

Huang Shan, Anhui Province, China

Pages 14, 52–53, 92–93

Katsuyama, Jigokudani, and Koshima Island, Japan

Pages 16–17, 70, 72, 128–29, 131, 144, 146–47, 162–63, 165

Kekopey Ranch, Kenya

Pages 12, 62, 80–81, 82, 112, 148–49

Mudumalai National Park, India

Pages viii, 61

San Diego Zoo, San Diego, California, United States

Pages 9, 18–19, 21, 25–26, 28–29, 32–33, 51, 57, 58–59, 64–65, 67, 79, 114, 123, 125, 126–27, 133, 152–53, 155, 156–57, 159, 170

Wisconsin Primate Center and Vilas Park Zoo, Madison, Wisconsin, United States

Pages iv, 40–41, 43, 44–45, 47, 89, 91, 107, 168

Yerkes Primate Center Field Station, Atlanta, Georgia, United States

Pages i, 13, 22–23, 30, 38–39, 49, 68–69, 76–77, 100–101, 103, 117, 121, 150–51, 166

List of Species in the Photographs

Bonobo

Pan paniscus

Brown or tufted capuchin monkey

Cebus apella

Chimpanzee

Pan troglodytes

Japanese (snow) monkey or macaque

Macaca fuscata

Lion-tailed macaque

Macaca silenus

Olive, or anubis, baboon

Papio cynocephalus anubis

Rhesus monkey or macaque

Macaca mulatta

Stump-tailed monkey or macaque

Macaca arctoides

Tibetan macaque

Macaca tibetana

Additional Photos

Page i: Infant brown capuchin monkey, about five days old (see p. 39).

Page iv: Newborn stump-tailed macaque with mother and other female onlookers (see p. 42).

Page vii: Other chimpanzees watching adult male Nikkie nibble on berries (see p. 136).

Page viii: Adult female lion-tailed macaque in southern India (see p. 60).

Page 166: Mother-infant pair of brown capuchin monkeys.

Page 168: Infant rhesus monkey.

Page 170: Juvenile male bonobo.

Sources Cited

Page 3: The Ansel Adams quotation is from *The Ansel Adams Guide: Basic Techniques of Photography*, by John P. Schaefer and Ansel Adams (Boston: Little, Brown and Co., 1999).

Page 127: The quotation about bonobos is from "Der afrikanische Anthropoide 'Bonobo': Eine neue Menschenaffengattung," by Eduard P. Tratz and Heinz Heck (*Saugetierkundliche Mitteilungen* 2 [1954]:97–101).

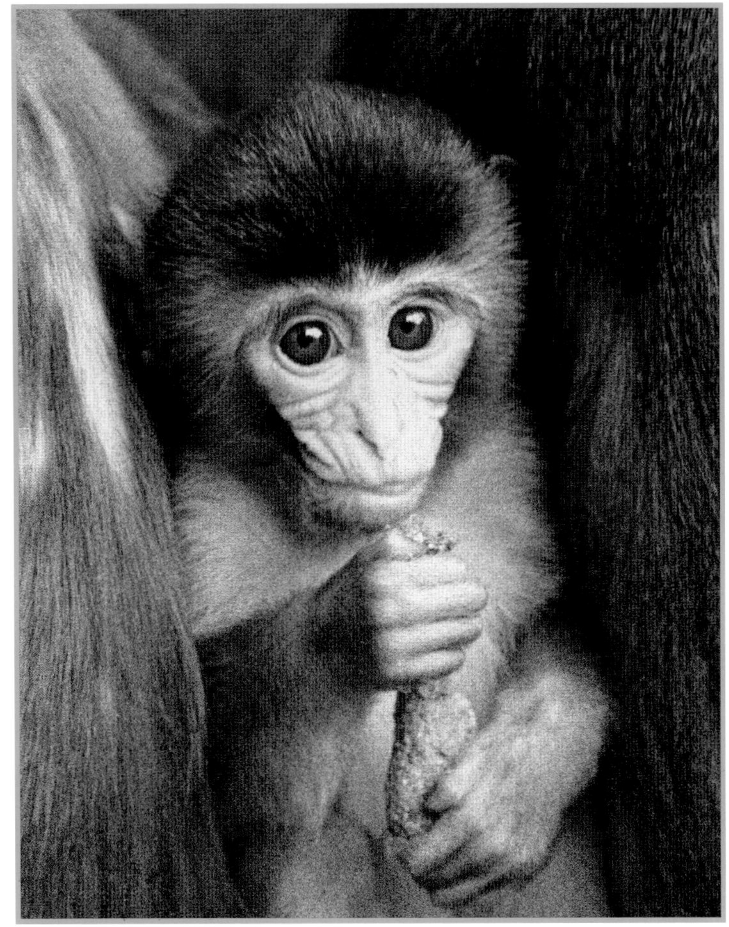

Acknowledgments

The photographs in this album have been taken at so many different locations that I cannot possibly thank every person who assisted me. I will limit myself to those colleagues directly relevant to my photography, such as Frank Kiernan and Bob Dodsworth. They headed the photography departments at the Yerkes Primate Center and Wisconsin Primate Center, respectively, and were always ready to develop my film, make contact prints, and offer advice about new technologies. I did build my own darkroom at home, after which I was able to do much of this work myself, but their reliable professional assistance has always been greatly appreciated.

For access to primates at locations away from my home base, I give heartfelt thanks to my colleagues Jinhua Li, Masayuki Nakamichi, Peggy O'Neill, Ronald and Betty Noë, Ren Mei Ren, Ananthakrishna Sharma, Mewa Singh, Karen Strier, Stephen Suomi, Ichirou Tanaka, and Shigeru Watanabe. For help in preparing this book, I wish to thank Darren Long for technical assistance and Marietta Dindo for high-resolution scanning directly from the negatives.

I am most grateful to my wife, Catherine Marin, who long ago taught me the basics of photography, has always been there for critical yet constructive feedback on my pictures, and has commented on every phase of the manuscript.

Finally, I feel indebted to my beloved subjects, who tolerated my persistent observations of them through an invasive camera eye. They must have wondered about this obsession of mine, undoubtedly taking it as one of those mysterious human quirks. My greatest reward was when they ignored me and went about their daily business.

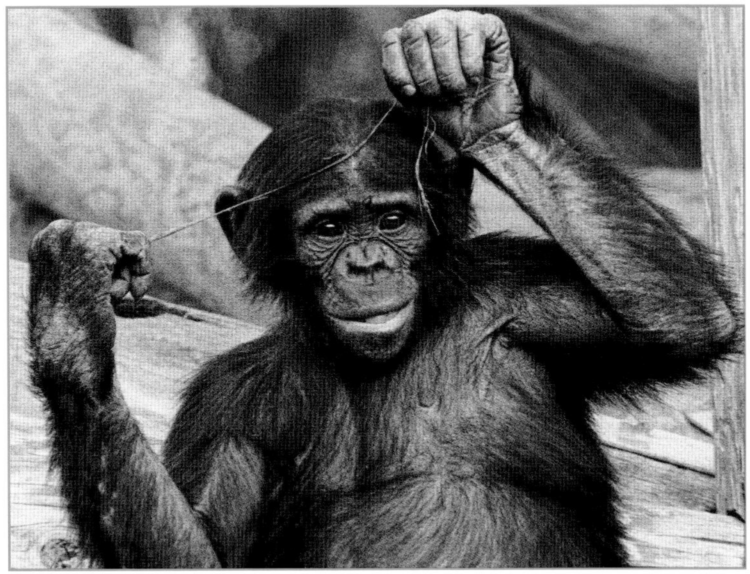

Recommended Reading

Readers who would like background information on primates and their behavior are referred to the following scientific and popular books.

The Primate Order

MacDonald, David. 1984. *The Encyclopedia of Mammals*. New York: Facts on File Publications.

Rowe, Noel. 1996. *The Pictorial Guide to the Living Primates*. New York: Pogonias.

Primate Behavior

de Waal, Frans B. M., ed. 2001. *Tree of Origin: What Primate Behavior Can Tell Us about Human Social Evolution*. Cambridge, Mass.: Harvard University Press.

Jolly, Allison. 1988. *The Evolution of Primate Behavior*. 2d ed. New York: Macmillan.

McGrew, William C., et al., eds. 1996. *Great Ape Societies*. Cambridge: Cambridge University Press.

Smuts, Barbara B., et al., eds. 1987. *Primate Societies*. Chicago: University of Chicago Press.

Strier, Karen B. 2003. *Primate Behavioral Ecology*. Boston: Allyn & Bacon.

Chimpanzees

de Waal, Frans B. M. (orig. 1982) 1998. *Chimpanzee Politics: Power and Sex among Apes*. Baltimore, Md.: Johns Hopkins University Press.

Goodall, Jane. 1986. *The Chimpanzees of Gombe*. Cambridge, Mass.: Harvard University Press.

Bonobos

de Waal, Frans B. M., and Frans Lanting. 1997. *Bonobo: The Forgotten Ape*. Berkeley: University of California Press.

Kano, Takayoshi. 1992. *The Last Ape*. Stanford, Calif.: Stanford University Press.

Primate Intelligence, Emotions, and Culture

Aureli, Filippo, and Frans B. M. de Waal, eds. 2000. *Natural Conflict Resolution*. Berkeley: University of California Press.

Byrne, Richard. 1995. *The Thinking Ape*. Oxford: Oxford University Press.

Cheney, Dorothy, and Robert Seyfarth. 1991. *How Monkeys See the World*. Chicago: University of Chicago Press.

de Waal, Frans B. M. 1989. *Peacemaking among Primates*. Cambridge, Mass.: Harvard University Press.

———. 1996. *Good Natured: The Origins of Right and Wrong in Humans and Other Animals*. Cambridge, Mass.: Harvard University Press.

———. 2001. *The Ape and the Sushi Master: Cultural Reflections by a Primatologist*. New York: Basic Books.

McGrew, William C. 1992. *Chimpanzee Material Culture*. Cambridge: Cambridge University Press, Cambridge.

Tomasello, Michael, and Josep Call. 1997. *Primate Cognition*. New York: Oxford University Press.